SHOW ME THE NUMBERS

Designing Tables and Graphs to Enlighten

STEPHEN FEW

Analytics Press
OAKLAND, CALIFORNIA

Analytics Press

PO Box 20313
Oakland CA 94620-0313
SAN 253-5602
www.analyticspress.com
Email: info@analyticspress.com

PUBLISHER: Jonathan G. Koomey

COPY EDITOR: Nan Wishner
PROOF EDITOR: Lisa Goldstein
COMPOSITION: BookMatters
COVER ART: Keith Stevenson
PHOTOGRAPHY: Frank Tapia
PRINTER AND BINDER: C&C Offset

ISBN: 0-9706019-9-9
ISBN-13: 978-09706019-9-5

Library of Congress Card Number 00-193242

This book was printed on acid-free paper in Hong Kong.

9 8 7

Heartfelt thanks to those fine human sparks here and there along the way who, through their kind words, thoughtful goading, and encouraging visions, taught me that I too could make a difference in the world.

Special gratitude also to my dear friends and former professional colleagues, Paul Winsberg and Diane Whitty, for their willingness to slog through the initial draft of this book, sacrificing many hours to give me valuable feedback and encouragement.

SHOW ME THE NUMBERS

Oh, the thirst to know
how many!
The hunger
to know
how many
stars in the sky!
We spent
our childhood counting
stones and plants, fingers and
toes, grains of sand, and teeth,
our youth we passed counting
petals and comets' tails.
We counted
colors, years,
lives, and kisses;
in the country,
oxen; by the sea,
the waves. Ships
became proliferating ciphers.
Numbers multiplied.
The cities
were thousands, millions,
wheat hundreds
of units that held
within them smaller numbers,

smaller than a single grain.
Time became a number.
Light was numbered
and no matter how it raced with sound
its velocity was 37.
Numbers surrounded us.
When we closed the door
at night, exhausted,
an 800 slipped
beneath the door
and crept with us into bed,
and in our dreams
4000s and 77s
pounded at our foreheads
with hammers and tongs.
5s
added to 5s
until they sank into the sea or madness,
until the sun greeted us with its zero
and we went running
to the office,
to the workshop,
to the factory,
to begin again the infinite
1 of each new day.

Excerpt from the poem
"Ode to Numbers"
by Pablo Neruda,
Selected Odes of Pablo Neruda,
translated by Margaret Sayers
Peden (1995), Berkeley:
University of California Press.

CONTENTS

12. THE INTERPLAY OF STANDARDS AND INNOVATION   237

When you design tables and graphs, you face a multitude of choices. Of the available alternatives, some are bad, some are good, some are best, and others are simply a matter of preference among equally good choices. By developing and following standards for the visual design of quantitative information, you can eliminate all but the best choices once and for all. Doing so dramatically reduces the time it takes to produce tables and graphs as well as the time required by your readers to make good use of them. Doing so sets your skills and creativity free to be used where they are most needed.

SHOW ME THE NUMBERS

1 INTRODUCTION

The use of tables and graphs to communicate quantitative information is common-place in business today, yet few of us who produce them have learned the design practices that make them effective. This introductory chapter prepares the way for a journey of discovery that will enable you to become an exception to this unfortunate norm.

Purpose
Scope
Intended readers
Content preview
Communication style

"Show me the numbers" is an expression that I've often heard during the course of business. Numbers are central to our understanding of business performance. They enable us to make informed decisions. The way we measure success in business is almost always based on numbers. We derive great value from the messages that these numbers convey, yet the significance of how we present them is rarely considered.

Contrary to popular wisdom, the data cannot always speak for themselves. Inattention to the design of quantitative communication results in a huge hidden cost to most businesses. Time is wasted struggling to understand the meaning and significance of the numbers—time that could be better spent doing something about them. What a shame, especially when you realize that this can be so easily remedied. To provide a practical solution to such a pervasive problem is the intention of this book.

Quantitative information—the numbers—takes us out of the realm of assumption, feeling, guesswork, gut instinct, intuition, and bias, into the realm of reliable fact based on measurable evidence. Too many business decisions are based on perceptions that are fallible. You may wake up in the morning, step outside, feel the sunshine on your skin, and know deep down in your bones that it's warmer today than it was yesterday, only to glance at the thermometer and discover that it is in fact five degrees cooler. More to the point of this book, your gut may tell you that business is now better than ever, but a careful check of the numbers could reveal that your market share has actually decreased during the past 12 months.

I don't intend to demean the value of well-honed business intuition. I personally have a strong intuitive sense of what's going on, of what will work and what won't, that is rarely wrong when applied to my areas of expertise. Rarely, but often enough to keep me humble. By trusting my gut—overly confident that the extra step of actually checking the numbers would be wasted—I have at times been dragged kicking and screaming to the embarrassing admission that I was wrong. The numbers have an important story to tell, and it is up to us to help them tell it.

Tables and graphs are generally the best means to communicate quantitative business information. They are so commonplace, many of us assume that knowledge of their effective use is common as well. *I assure you, it is not.* Evidence of this fact in the form of countless poorly designed tables and graphs is everywhere. According to Karen A. Schriver, an internationally recognized expert in document design, "Poor documents are so commonplace that deciphering bad writing and bad visual design have become part of the coping skills needed to navigate in the so-called information age."[1]

1. Karen A. Schriver (1997) *Dynamics in Document Design.* New York: John Wiley & Sons, Inc., page xxiii.

The reason for this sad state of affairs is simple—very few of us have ever been trained to design tables and graphs effectively. Some of us have struggled to do this work well but simply haven't found useful resources to assist us. Others of us haven't struggled at all, simply because we haven't seen enough examples of good design to recognize the inadequacy of our own efforts.

Before saying anything more, let me show you what I mean. Take a moment to examine the graph below.

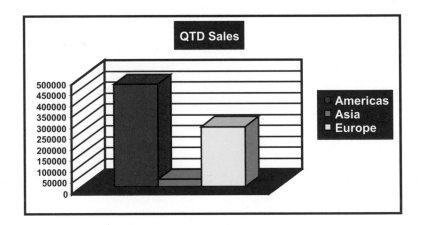

FIGURE 1.1 This is a typical example of a poorly designed business graph. Notice the attempt at artistic flair in the use of color, 3-D, and shading of the vertical bars.

This graph has visual impact—it's dramatic, it's colorful, it jumps off the page and demands attention—but to what end? What's the message?

Let's evaluate its effectiveness in light of its objective. Assume that its purpose is to inform a corporation's executives every Monday morning about the current state of quarterly sales, split into three geographical regions: Americas, Europe, and Asia. Given this intention, look at it again. What message do the executives get from this graph? Put yourself in their position. Take a minute or two to examine the graph and interpret its message in light of your executive interests.

. . . .

What did you get? Probably something like the following:

* Sales for the Americas are better than sales for Europe, which in turn are better than sales for Asia. This much is clear.

- Sales for Asia don't amount to very much compared to the other regions.
- Sales for the Americas are more than 400,000, or thereabouts. Sales for Europe are somewhere around 200,000. Sales for Asia appear to be around 50,000, perhaps a little less.

That's not much information, and you certainly had to work for it, didn't you? Here are the actual values that were used to create this graph:

- Americas = $469,384
- Europe = $273,854
- Asia = $34,847

As an executive, you may not need to know the precise sales amounts, but I suspect you would want better accuracy than you can discern from this graph, and you would certainly want to get it faster and with far less effort.

Several pieces of critical information aren't supplied by this graph:

- Given the fact that these sales are international, what is the unit of measure? Is it U.S. dollars, Euros, etc.?
- Quarter-to-date sales as of what date? If you filed this report away for future reference and pulled it out again a year from now, you wouldn't know what day, what quarter, or even what year it represents.
- How do these sales figures compare to your plan for the quarter?
- How do these sales figures compare to how you did at this time last quarter or this same quarter last year?

This graph lacks important contextual information and critical points of comparison. As a report of quarter-to-date sales across your major geographical regions, it doesn't show you very much. It uses a great deal of ink to say very little.

Given the intended message and the information that an executive might find useful, the following display tells the story better.

2003 Q1-to-Date Regional Sales
March 15, 2003

	Sales (U.S. $)	Percent of Total Sales	Current Percent of Qtr Plan	Projected Sales (U.S. $)	Qtr End Projected Percent of Qtr Plan
Americas	469,384	60%	85%	586,730	107%
Europe	273,854	35%	91%	353,272	118%
Asia	34,847	5%	50%	43,210	62%
	$778,085	100%	85%	$983,212	107%

Note: To date, 83% of the quarter has elapsed.

FIGURE 1.2 This table contains all of the information that was contained in the graph in *Figure 1.1*, plus much, much more, without becoming complicated.

This is a simple table, but it's easy to read, and it contains a great deal more information. No ink is wasted. As an executive, you could use this report to make important decisions. It communicates.

Here's another fairly typical example of a graph that suffers from severe design problems:

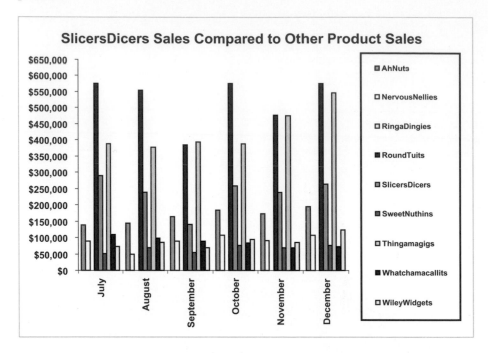

FIGURE 1.3 This is a typical example of a poorly designed vertical bar graph.

Without the graph's title, would you have any idea that its purpose is to compare the sales performance of the product named SlicersDicers to the performance of each of the other products? In the general field of design, we speak of things having *affordances*—characteristics that reveal how they're to be used. A teapot has a handle. A door that you need to push has a push-plate. The design of an object should, in and of itself, suggest how the object should be used. This graph relies entirely on its title to declare how the graph should be used. Not only does the design fail to suggest the graph's use, the design actually subverts its use.

For a thorough examination of *affordances* in the broader context of design for usability, see Donald A. Norman's classic text, *The Design of Everyday Things* (1988). New York: Basic Books.

Imagine for a few minutes that it is your job to assist the manager responsible for the products that appear in this graph. You've been asked to create a simple means to help her compare the performance of SlicersDicers to the performance of other products in her portfolio. For her immediate purposes, she isn't particularly interested in its sales through time or in actual dollars. She just wants to know how sales of SlicersDicers compare to those of her other products. To begin your task, take some time to examine the graph again; then, make a list of the ways its design fails to support the product manager's needs. Take a minute to write your list next to the graph in the right margin of the book.

Some of my friends cringe at the thought of actually encouraging readers to write in a book. Nevertheless, I strongly encourage the practice as a means to keep notes and highlight important content where it will be most useful in the future. I do discourage this practice, however, when the book is borrowed.

· · · · · · ·

Now, advance this exercise to the next level. Take a few more minutes to list the features you would incorporate into a new version of the graph to better serve the product manager's needs. Make use of your prior list to suggest areas that you should address.

· · · · · · ·

How did you do? Did you find yourself rising to the challenge? If so, you are already beginning to think critically and creatively about the design of tables and graphs even before we've begun to examine specific principles of good design. There is no single correct solution to the task at hand, but let me show you one possible solution that previews a design technique that may interest you.

FIGURE 1.4 This is a series of related graphs, each designed to compare the sales of the SlicersDicers product to those of a different product.

Given the goal of communicating how SlicersDicers compares to each of the other products, can you see the ways that this design incorporates specific affordances to support this goal? Take a minute longer to identify as many of these features as you can.

· · · · · · ·

One powerful affordance is the reference line of 100% representing the sales of SlicersDicers in each of the graphs, which makes crystal clear that the purpose is to compare the sales of each of the other products to those of SlicersDicers.

Let's look at one last example, for now, of ineffective graph design. Everyone is familiar with the ubiquitous pie chart. This one is quite typical in its design:

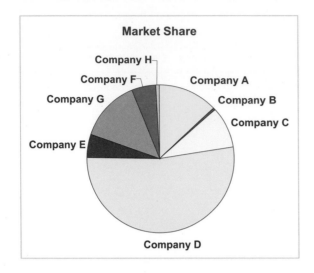

FIGURE 1.5 This is an example of an ineffective style of graph: a pie chart.

Here's the same graph below, but this time it's dressed up a little with the addition of 3-D:

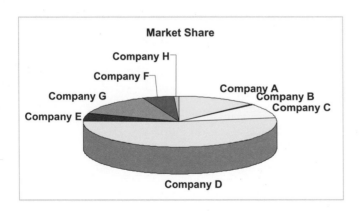

FIGURE 1.6 This is a slight variation of *Figure 1.5*, this time incorporating a 3-D perspective. This design is even less effective.

Did the use of 3-D enhance the display? These two graphs are meant to display the same market share numbers, yet notice how the addition of 3-D perspective affects your perception of the data. The intention of these graphs is for Company G to compare its market share to the shares of its competitors. Can you determine Company G's market share from either of these graphs? Can you determine its rank compared to its competitors? Which has the greater share: Company A or Company G? Because of fundamental limitations in visual perception, you really can't answer any of these questions accurately.

Now look at the exact same market share numbers displayed in a more effective manner:

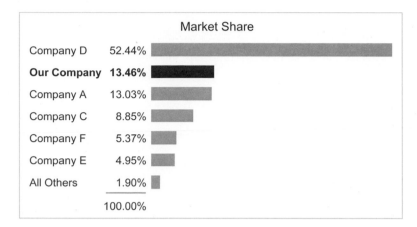

FIGURE 1.7 Percentages can be displayed in a number of ways. This example uses a traditional bar graph (a.k.a. horizontal bar) to do the job.

Did you have any trouble interpreting this information? Did you struggle to find the most important information? It's obvious that Our Company ranks second, slightly better than Company A, and that its market share is precisely 13.46%. This display contains no distractions. It gives the important numbers clear voices to tell their story. This communicator did a good job.

Purpose

Despite their commonplace use today, graphs have only been employed to display quantitative information for a few centuries, and, even though tables have been around longer, it wasn't until the last quarter of the 20th century that the use of either became widespread. What caused this rise in their use? The *personal computer*.

Shortly after the advent of the PC, I began developing and teaching courses in the use of some of the earliest PC business software, including *Lotus 1-2-3*. Although Lotus wasn't the first electronic spreadsheet product, and it's no longer the most popular electronic spreadsheet, it legitimized the PC as a viable tool for business. Prior to the advent of spreadsheet software, tables of quantitative information were generally produced using pencils or pens, a sheet of lined paper, a calculator, and hours of tedious labor. Graphs could only be produced using a pen or pencil (perhaps several of different colors), a straight-edged device (e.g., a ruler or draftsman's triangle), a sheet of graph paper, and, once again, hours of meticulous labor. When chart-producing software hit the scene, however, many of us who would have never before taken the time to draw a graph suddenly became Rembrandts of the X and Y axes—or so we thought. Like kids in a toy store, we went crazy over all the available colors and cool effects, thrilled with the new means for techno-artistic expression. Through the *magic* of computers, making tables and graphs became easy—perhaps too easy.

Today, everyone can produce reports of quantitative business information in the form of tables and graphs. Children in elementary school are taught the mechanics of using spreadsheet software. Something produced with a computer, however, acquires an air of authenticity and quality that it doesn't necessarily deserve. In our excitement to produce what we could only make before with great effort, many of us have lost sight of the real purpose of quantitative displays—*to provide the reader with important, meaningful, and useful insight.* To communicate quantitative information effectively first requires an understanding of the numbers, then the ability to display their message for accurate and efficient interpretation by the reader.

The necessary knowledge and skills are well within your reach if you make a little effort. Once you've read this book and practiced a bit, you'll find that it is no more difficult and takes no longer to produce effective tables and graphs than it does to produce ineffective ones. By synthesizing the best practices of quantitative information design that have been learned through many years of research and real-world trial and error by pioneers who have blazed the trail, I hope to make your effort relatively easy, and perhaps even enjoyable.

The purpose of quantitative tables and graphs in business is to communicate important information effectively. That's it. Not to entertain, not to indulge in self-expression, not to make numbers interesting through flash-and-dazzle that you would otherwise deem boring. Edward Tufte, the best-known expert in this field, expresses this perspective quite simply: "The overwhelming fact of data graphics is that they stand or fall on their content, gracefully displayed . . . Above all else show the data." And to those who feel the need to dress up the numbers, he warns: "If the statistics are boring, then you've got the wrong numbers."[2]

As with any endeavor that is worthwhile, our goal in the design of quantitative business information for the purpose of communication is *excellence*. Why anything less when the means to attain excellence are readily available? Why anything less when doing *good work* is so satisfying, and simply *getting by* is such a drag? With the right knowledge and skills, you have a chance to make a difference. Even if you're not inspired by your employer's objectives, do it for yourself. Good work, or what the Buddhists call "right livelihood," is one of fundamental delights of life. It is fulfilling in and of itself.

As providers of quantitative business information, it is our responsibility to do more than sift through the data and pass it on; we must help our readers gain the insight contained therein. We must design the message in a way that leads readers on a journey of discovery, making sure that what's important is clearly seen and understood. The right numbers have an important story to tell. They rely on you to give them a clear and convincing voice.

Scope

One of the challenges in writing a book like this is to constrain its scope. The contents must compose a coherent whole that addresses a specific set of closely related needs for a well-defined audience. There is a risk of communicating too little by saying too much. This is indeed a challenge when your interests are far ranging, like mine. What to leave out? In light of this objective, I've limited this book to content that is 1) practical and applicable to typical business reporting, 2) focused on communication, and 3) focused on design.

All of the tables and graphs in this book were created using *Microsoft Excel* software, in part to demonstrate that good design can be achieved using software that is familiar to most business users of computers.

2. Edward R. Tufte (1983) *The Visual Display of Quantitative Information.* Cheshire CT: Graphics Press. The first quotation appears on pages 121 and 92; the second on page 80.

Although some businesses employ professional statisticians to perform sophisticated quantitative analyses, most of us never touch sophisticated statistics and would be intimidated by any statistic more complicated than an average. Although every business has a few custom measures and perhaps one or two particular display techniques that fit its specialized nature, measures and display techniques that are not of interest to most of us do not appear in this book. Rest assured that you will find little in this book that doesn't apply directly to the work that you do at your own place of business. Furthermore, you will not have to wade through information that, however interesting from an academic perspective, is peripheral to the task at hand.

Tables and graphs of quantitative business information can be used for four purposes:

- Analyzing
- Monitoring
- Planning
- Communicating

When you use tables and graphs to discover the message in the data, you are performing analysis. When you use them to track information about your internal performance, such as the speed or quality of manufacturing, you are engaged in monitoring. When you use them to prepare for the future, such as in budgeting, you are planning. When you use them to pass on to others a message about the business, however, your purpose is communication, no matter what the content. All of these are important uses of tables and graphs, but the process that you engage in and the design principles that you follow differ for each. My purpose in this book is to help you learn to design tables and graphs to communicate important business information.

This book focuses on design. It is not a book about the mechanics of constructing tables and graphs using a particular software product (e.g., *Microsoft Excel*). It is not an encyclopedia of tables and graphs, listing and describing the countless variations that exist. It is not a book that will make you an expert in a particular type of graph, such as in the nuances of histograms. It is not a book about making tables and graphs look pretty though there is certainly beauty in those that simply and elegantly hit their communication target dead on. It is a book about design, in particular about design practices that apply broadly to using tables and graphs for effective quantitative business communication.

When do you use a table versus a graph? When do you use one type of graph rather than another? How do you highlight what's most important and make the message crystal clear? What do you avoid in order to eliminate anything that might distract from that message? What you will learn from this book can be applied across the board to every table and graph you will ever need to create.

Intended Readers

Simply stated, this book is intended for anyone whose work involves the creation and use of tables and graphs for the communication of quantitative business information. Those of us with such responsibilities don't fit into a tidy set of job titles. Our roles are spread across a broad spectrum. Some of us serve in positions that specialize in the production of tables and graphs, with job titles that often include the term *analyst*, such as *Financial Analyst*, *Business Analyst*, *Data Analyst*, or *Decision Support Analyst*, to name a few. Some of us have managerial responsibilities and are required occasionally to prepare tables and graphs for other managers higher up in the ranks. Some of us are graphic artists, and someone decided that because the words *graphs* and *graphic* are related, responsibility for the production of quantitative graphs must fall into our area. The rest of us are spread all over the organization chart. No matter what it says on your business card, if you are responsible for creating tables and graphs to communicate quantitative information, and you want to do it well, this book is for you.

Content Preview

Tables and graphs don't just display numbers; they present them in a manner that relates them to something, such as to time or to one another, in order to reveal a meaningful message. Tables and graphs are two members of a larger family of display methods known as *charts*. In addition to tables and graphs, there are other types of charts, such as diagrams, which illustrate a process or set of relationships, and maps, which depict phenomena spatially.

Although tables and graphs are both vehicles for presenting information visually, they differ significantly in the role that visual perception plays in reading and interpreting their information. Graphs are perceived almost entirely by our *visual* system, and, as such, employ a visual language of sorts. Lines, bars, and a variety of other symbols, positioned within a two-dimensional space formed by axes, are used to communicate visual patterns and relationships. To see patterns and relationships is a natural function of visual perception.

Tables, with their columns and rows of information, interact primarily with our *verbal* system, which entails what we normally mean when we speak of language. Thus, we process the information in tables in a sequential fashion, reading down columns or across rows of numbers, comparing this number to that number, one pair at a time. This is different from the visual processing that occurs when viewing graphs, which involves high-bandwidth, simultaneous input of multiple data, enabling us to perceive a great deal of quantitative information in a burst of recognition. One method of quantitative display is not better than the other, but each is better than the other for particular communication tasks, and both play a vital role in business communication.

This book is designed to take you on a journey. We begin in Chapter 2, *Numbers Worth Knowing*, with an introduction to a few numbers that are quite handy and particularly useful in tables and graphs. Numbers are the message bearers and the protagonists of this book, so it is natural that we start with them.

We continue in Chapter 3, *Fundamental Concepts of Tables and Graphs*, with an introductory examination of tables and graphs, including what is common to both, then proceeding to which works best for what purposes.

In Chapter 4, *Fundamental Variations of Tables*, we go on to break tables down into the types of information they can be used to display and into the basic ways they can be structured.

In Chapter 5, *Fundamental Variations of Graphs*, we likewise break graphs down into the types of information they can be used to display and explore the ways that data can be visually encoded to present that information most effectively.

With that foundation in place, in Chapter 6, *Visual Perception and Quantitative Communication*, we take a detour to learn how visual perception works, from the point when light enters our eyes to the point when the information that our brains have gleaned from the light is stored for future reference in memory or is discarded. You may be tempted to skip this chapter. Don't give in. If you want to understand what works and what doesn't in the design of graphs so that you can apply your knowledge to each new situation that arises, you will need to understand a few fundamental concepts about visual perception.

With this knowledge of visual perception under our belts, we proceed in Chapter 7, *General Design for Communication*, to apply that knowledge to visual design in the form of general practices for both tables and graphs.

From there, we dig into the details of *Table Design* in Chapter 8. We examine the structural components of tables, how to combine them for optimal effect, and the all-too-common erroneous practices that you should avoid.

We then shift our focus in Chapter 9, to *General Graph Design*, beginning with design principles that apply to all types of graphs. In Chapter 10, *Component-Level Graph Design*, we look closely at each component of graphs to learn when and how to use them for effective communication. We complete our examination of graphs with Chapter 11, *Design Solutions for Multiple Variables*, where we develop some useful strategies for displaying complex messages that consist of several sets of data.

We end our journey in Chapter 12, *The Interplay of Standards and Innovation*, where I climb up on my soapbox and proclaim the inevitable pain and suffering that you will experience if you fail to establish and follow a set of standards for the design of tables and graphs.

Communication Style

At heart, I am a teacher. My mind works like a teacher's. When I tell you about something, I care that you get it. When you do, I feel good. When you don't, I feel that I've failed. Consequently, this book is designed as a learning experience, not simply to inform or entertain. It is not designed as a reference that you pull from the shelf occasionally. It is designed to get into your head in a way that is thorough and lasting.

As a result, it is filled with examples that bring the material to life, as well as questions that invite you to think and perhaps even struggle a bit during the process. It contains exercises that give you an opportunity to practice what you're learning and to learn more thoroughly through that practice. It is laid out in a sequence that leads you through learning at a conceptual level, then allows you to apply that conceptual understanding to various real-world scenarios. I take you on a journey of discovery, rather than presenting principles that you must memorize and follow based merely on my authority as an expert. I want you to learn these practices and to make them a part of your work to the point where you no longer need to think about them. If someone asks you why you do what you do five years from now, I want you to still be able to explain it.

I love teaching, in part because I love learning. I try to approach each day as a student. Doing so enriches me and keeps life very interesting. I invite you to approach this book with the curiosity of a student. If you do, the rewards will be worth the effort.

2 NUMBERS WORTH KNOWING

Quantitative information forms the core of what businesses must know to operate effectively. The current emphasis on business metrics, Key Performance Indicators (KPIs), and Balanced Scorecards demonstrates the importance of numbers in business. The messages contained in numbers are communicated most effectively when you understand the fundamental characteristics and meaning of the numbers that are commonly used in business, as well as the fundamental principles of effective communication that apply specifically to quantitative information.

> **Quantitative relationships**
> > **Relationships within categories**
> > **Relationships between quantities**
> **Numbers that summarize**
> > **Measures of average**
> > **Measures of distribution**
> > **Measures of correlation**
> > **Measures of ratio**
> **Measures of money**

Numbers are not intrinsically boring. Neither are they intrinsically interesting. The fact that they are quantitative has no bearing on their inherent appeal. They simply belong to the class of information that communicates the quantity of something. The impact and appeal of information, quantitative or not, flows naturally from the significance and relevance of the message it contains. As a communicator, it is up to you to give a clear and unobstructed voice to that information and its message, using language that is easily understood by your audience.

You may be anxious to jump right into the design of tables and graphs. After all, that's the fun stuff. I must admit, I was tempted to get right to it, but because numbers are the content and thus the substance of tables and graphs, it's important to begin our journey by getting acquainted with a few numbers that are particularly useful in quantitative communication.

Quantitative Relationships

When you design the display of quantitative information, whether you use a table or graph, the specific type of table or graph you use depends primarily on your message. What about the message? Quantitative messages are always about *relationships*. Numbers, in and of themselves, are of no use unless they measure something that is important to you. Here are some common examples of relationships that define the essential nature of quantitative messages:

Quantitative Information	Relationship
Units of a product sold per geographical region	Sales related to geography
Revenue by quarter	Revenue related to time
Expenses by department and month	Expenses related to organizational structure and time
A company's market share compared to that of its competitors	Market share related to companies
The number of employees who received each of the five possible performance ratings (1–5) during the last annual performance review	Employee counts related to performance ratings

In each of these examples, there is a simple relationship between some measure of quantity and one or more associated categories of interest to the business (geography, time, etc.). Quantitative information consists of two types of data: *quantitative* and *categorical*. Quantitative values measure things; categories subdivide the things that they measure into useful groups, such as geographical areas (e.g., north, east, south, and west) in the category called sales regions, or individual months in the category called time. This distinction between quantitative and categorical data is fundamental to tables and graphs. These two types of data play different roles in tables and graphs and are often structured and displayed in distinct ways.

Quantitative values are also expressed in units of measure. For instance, the quantitative value *$200* is made up of the quantity—200—and a relevant unit of measure—dollars.

Sometimes the relationships we display are simple associations between quantitative values and categorical subdivisions, such as those in the previous examples. Sometimes the relationships display direct associations between multiple sets of quantitative values, such as in the examples below:

Quantitative Information	Relationship
The effect of a mass-mailing marketing campaign on order volume	The number of letters sent related to the number of orders received
Units sold and the resulting revenue in correlation to pricing	Product price related to the associated number of units sold

This distinction between simple relationships that associate quantitative values and categorical subdivisions and somewhat more complex relationships that associate multiple sets of quantitative values is also fundamental to our use of tables and graphs. Different types of relationships require different types of displays.

So far we've only examined a few examples of quantitative relationships, but the list is endless. Think for a minute or two about the quantitative information that is communicated at your place of business. Can you think of any that doesn't involve relationships?

Thus far we've learned the following about quantitative information:
- Quantitative information consists of two types of data:
  - Quantitative
  - Categorical
- Quantitative information always describes relationships.
- These relationships involve either
  - Simple associations between quantitative values and categorical subdivisions or
  - More complex associations among multiple sets of quantitative values.

In addition to the two fundamental types of quantitative relationships that we've already noted, there are also a variety of ways in which categorical subdivisions or the quantitative values associated with them can relate to one another. Let's take a look at these ways.

Relationships Within Categories

Categorical subdivisions can relate to one another in the following ways:

- Nominal
- Ordinal
- Interval
- Hierarchical

NOMINAL

A *nominal* relationship is one in which the individual subdivisions of a single category are discrete and have no intrinsic order. For instance, the four sales regions *East*, *West*, *North*, and *South*, in and of themselves, are not related in any particular order. These labels simply name the different sales regions, thus the term nominal, which means "in name only." Here's a simple example:

Region	Sales
North	139,883
East	135,334
South	113,939
West	188,334
Total	$577,490

FIGURE 2.1 This is an example of a nominal relationship.

When you communicate a quantitative message that is nominal in nature, you simply divide up the quantitative values in association with separate categorical subdivisions, each bearing a different name, but your message does not relate those subdivisions to one another in any particular way.

ORDINAL

An *ordinal* relationship between categorical subdivisions is one in which the individual subdivisions have a prescribed *order*. Typical examples include "first, second, third . . ." and "small, medium, and large." To display them in any other order would rarely be meaningful.

INTERVAL

An *interval* relationship is one in which the categorical subdivisions consist of a series of individual, sequential numerical ranges that subdivide a full set of quantitative values into smaller ranges. These individual numerical ranges, called intervals, can be arranged in order from smallest to largest (ascending order) or largest to smallest (descending order). Interval relationships are used when you wish to look at how something is distributed across a broad range of quantitative values by subdividing the range into a set of smaller, more manageable ranges. Here's a common example:

Order Size (U.S. Dollars)	Order Quantity	Order Amount
>= 0 and < 1,000	17,303	6,688,467
>= 1,000 and < 2,000	15,393	26,117,231
>= 2,000 and < 3,000	10,399	29,032,883
>= 3,000 and < 4,000	2,093	6,922,416
>= 4,000 and < 5,000	1,364	5,805,184
Total	46,552	$74,566,181

FIGURE 2.2 This is an example of an interval relationship. Notice that the intervals are equal in size. This is especially important when you intend to graphically display the distribution of a set of values across a range, called a *frequency distribution*.

In this example, to see how the orders were distributed across the entire range of order sizes, it wouldn't make sense to count the number of orders and sum their totals for each individual order amount, because that would involve an unmanageably large set of order sizes. The solution involves subdividing the full range of order sizes into a series of contiguous ranges.

Take a moment to test what you've learned so far. Look at the example below and determine which of the three relationships—nominal, ordinal, or interval—best describes its categorical subdivisions of time (months in this case).

Dept	Jan	Feb	Mar	Q1 Total
Marketing	83,833	93,883	95,939	273,655
Sales	38,838	39,848	39,488	118,174
HR	37,463	37,939	37,483	112,885
Finance	13,303	14,303	15,303	42,909
Total	$173,437	$185,973	$188,213	$547,613

FIGURE 2.3 This is an example of time-series relationship.

Your initial inclination was probably to conclude that categorical subdivisions of time are ordinal, for they certainly make sense only when arranged chronologically. This begs the further question, however, "Do these subdivisions of time represent intervals along a quantitative scale?" The answer is "Yes, they do." Time is a quantitative scale that measures duration. Even though different months do not all represent the same exact number of days and are therefore not precisely equal intervals, for reporting purposes we treat them as equal.

So far the categorical relationships that we've examined involve relationships between members of the same categorical set. The remaining relationship discussed below does not.

HIERARCHICAL

A *hierarchical* relationship involves multiple categories that are closely related to each other as separate levels in a ranked arrangement. Starting from the top of the hierarchy and progressing down, each subdivision at each level is associated with only one subdivision at the level above it. Each subdivision at every level, except the bottom level, can have one or more subdivisions associated with it in the next level down. This is much easier to show than to describe with words. Here's a typical example viewed from left to right:

Division	Dept	Group	Expenses
G&A	Human Resources	Recruiting	42,292
		Compensation	118,174
	Info Systems	Operations	512,885
		Applications	442,909
Finance	Accounting	AP	79,000
		AR	55,542
	Corp Finance	Fin Planning	93,027
		Fin Reporting	74,383

FIGURE 2.4 This is an example of a hierarchical relationship. The *G&A* division is composed of two departments: *Human Resources* and *Info Systems*. The *Recruiting* and *Compensation* groups belong to the *Human Resources* department, and the *Operations* and *Applications* groups belong to the *Info Systems* department.

Hierarchical relationships between categories are commonly used in tables, and, to a lesser degree, in graphs, to organize quantitative information.

Relationships Between Quantities

Categorical subdivisions can also relate to one another by virtue of the quantitative values associated with them. The quantitative values can be used to display the following relationships:

- Ranking
- Ratio
- Correlation

RANKING

When the order in which the categorical subdivisions are displayed is based on the associated quantitative values, either in ascending order or descending order, the relationship is called a *ranking*. If you need to construct a list of your company's top five sales orders for the current quarter based on revenue, the message would be enhanced if you arranged them by size, in this case from the largest to the smallest of the five, as you see in the following figure:

Technically, the term *ordinal* could be used to describe a ranking relationship as well, but I'm using distinct terms to highlight the difference between a sequence based on categorical subdivisions and one based on quantitative values.

Rank	Order Number	Order Amount
1	100303	1,939,393
2	100374	875,203
3	100482	99,303
4	100310	87,393
5	100398	67,939
		$3,069,231

FIGURE 2.5 This is an example of a ranking relationship.

RATIO

A *ratio* is a relationship in which two quantitative values are compared by dividing one by the other. This produces a number that expresses their relative quantities. A common example is the relationship of the quantitative value for a single categorical subdivision compared to the sum of the entire set of subdivisions in the category (e.g., the sales of one region compared to the total sales of all regions). The ratio of a part to its whole is generally expressed as a percentage where the whole equals 100%, and the part equals some lesser percentage. Here's an example of a part-to-whole ratio in tabular form, which displays market share information for five companies, both in actual dollar sales and in percent-of-total sales:

Company	Sales	Sales %
Company A	239,949,993	15%
Company B	873,777,473	54%
Company C	37,736,336	2%
Company D	63,874,773	4%
Company E	399,399,948	24%
Total	$1,614,738,523	100%

FIGURE 2.6: This is an example of a part-to-whole ratio.

When you want to compare the size of one part to another or to the whole, it is easier, more to the point, and certainly more efficient for your audience to interpret a table or graph that contains values expressed as percentages. This is true because percentages provide a common denominator, a common frame of reference—not just any common denominator but one with the nice round value of 100, which makes comparisons very easy to understand.

Another common use of ratios in business involves measures of change. When the value of something is tracked through time, it is often useful to note how it changes from one point in time to the next. Here's a common example of a ratio used to express change, in this case change in expenses from one month to the next:

| Department | Expenses | | | |
	Jan	Feb	Variance	Change %
Sales	9,933	9,293	-640	-6%
Marketing	5,385	5,832	+447	+8%
Operations	8,375	7,937	-438	-5%
Total	$23,693	$23,062	-$1,327	-3%

FIGURE 2.7 This is an example of a ratio used to compare the expenses from one month to the next.

CORRELATION

A *correlation* is a relationship in which the values of two paired sets of quantities are compared to determine whether increases in one set correspond to either increases or decreases in the other set. For instance, is there a correlation between the number of years employees have been doing particular jobs and their productivity in those jobs? Does productivity increase along with tenure, does it decrease, or is there no significant correlation in either direction?

Thus far we've learned the following about quantitative information:

- Quantitative information consists of two types of data:
  - Quantitative
  - Categorical
- Quantitative information always describes relationships.
- These relationships involve either
  - Simple associations between quantitative values and categorical subdivisions or
  - More complex associations among multiple sets of quantitative values.
- There are four types of relationships within categories:
  - Nominal
  - Ordinal
  - Interval
  - Hierarchical
- There are three types of relationships between quantitative values:
  - Ranking
  - Ratio
  - Proportion

We have not covered a comprehensive list of possible quantitative relationships. Rather, we've homed in on those that are most relevant to the common uses of numbers in business. If you're wondering why these different types of quantitative relationships are important enough to cover in this chapter, hold on for a while. When we get to the later chapters on tables and graphs, the significance of these quantitative relationships and your ability to identify them will become clear. You'll discover that there are many distinct table and graph design methods and principles that connect directly to these different quantitative relationships.

Numbers that Summarize

Now for some basic *business statistics*. This is liable to be one of the briefest and simplest looks at statistics that you'll ever encounter. The truth is, most of us who communicate quantitative business information can get by with a limited statistical vocabulary.

Statistics provide several methods for data reduction—in other words, *summarization*, or what we sometimes call *aggregation*. Often, your quantitative message is best communicated by reducing large sets of numbers to a few numbers, allowing your readers to easily and efficiently comprehend and assimilate the message the numbers convey. If an executive asks you how sales are doing this quarter, you wouldn't give her a report that listed each individual sales order; you would give her the information in summary form. Relevant data might include such aggregates as the *sum* of sales orders in U.S. dollars, the *count* of sales orders, and perhaps even the *average* sales order size in U.S. dollars.

We have several ways to summarize numbers, some of which are visual in nature and apply only to graphs, which we'll thoroughly explore later, and some of which are purely statistical in nature, which is our focus in this chapter. Summing and counting sets of numbers are the most common means of aggregation used in business-related quantitative communication. Because I assume that you already understand counts and sums, we'll skip them and proceed directly to the other more complex data reduction methods that are particularly useful in business.

Measures of Average

Let's begin with a question. Take a moment to finish this sentence: "An average represents . . ."

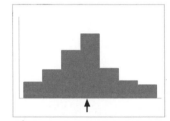

. . . .

It's interesting how many terms we carry around in our heads and use without ever really knowing how to define them. Ever had a child ask you what something quite familiar means and found yourself struggling for adequate words? If the concept of an average is one of those terms for you, here's a definition:

> An average is a single number that represents the middle of an entire set of numbers.

There are actually four distinct entities that are used in statistics to measure the middle of a set of quantitative values, and all of them are called averages:

- Mean
- Median
- Mode
- Midrange

It's useful to understand how these four differ, for they are each designed to work best in particular circumstances. Selecting the wrong type of average for your message could result in misleading information.

MEAN

Normally, when most of us think of an average, we think of what is more precisely called the *arithmetic mean* or simply the *mean*. In fact, many software products label the function that calculates the mean as "Average" (or sometimes "AVG" for short). Statisticians must cringe when they see this. Statistical software wouldn't make this mistake. Means are calculated as follows:

> Sum all the values; then, divide the result by the number of values.

Here's an example:

Quarter	Units Sold
Q1	339
Q2	373
Q3	437
Q4	563
Sum	1,712
Count	4
Mean (per Qtr)	428

FIGURE 2.8 This is an example of a *mean*, calculated as 1,712 (the sum) divided by 4 (the count), equaling 428.

A mean is the simplest type of average to calculate (excluding the *midrange*, which is seldom used), and the type most commonly supported by software. However, the mean isn't always the best choice for your message.

Means provide a measure of the middle in a manner that takes every value into account, no matter how extreme. Sometimes this is exactly what you need, but sometimes not. Take a look at the following example, and see if you can determine why use of the mean would be a misleading summary of employee salaries in the marketing department if your intention is to express the *typical* salary.

Employee	Position	Annual Salary
Employee A	Vice President	475,000
Employee B	Manager	165,000
Employee C	Manager	165,000
Employee D	Admin Assistant	43,000
Employee E	Admin Assistant	39,000
Employee F	Analyst	65,000
Employee G	Analyst	63,000
Employee H	Writer	54,000
Employee I	Writer	52,000
Employee J	Graphic Artist	64,000
Employee K	Graphic Artist	62,000
Employee L	Intern	28,000
Employee M	Intern	25,000
	Mean Salary	$100,000

FIGURE 2.9 This is an example of the use of a statistical mean in circumstances for which it is not well suited.

Why doesn't the mean work well for this purpose? The mean in this case is skewed heavily toward the higher salaries, giving the impression that employees are typically better compensated than they actually are. What you're seeing here is the fact that the mean is very sensitive to extremes. The Vice President's salary is definitely an extreme, a value that falls far outside the norm. When you need a measure that represents what is typical of a set of values, you would want to use an average that is not so sensitive to extremes.

Statisticians refer to extreme values in a data set (i.e., those that are located far away from most of the values) as *outliers*. The Vice President's salary in *Figure 2.9* is an outlier.

MEDIAN

The statistical *median* is the average that comes in handy when you need to communicate quantitative messages such as the one in the above example because the median is not at all sensitive to extreme values.

Medians are calculated as follows:

> **Sort the values in order** (either high to low or low to high); then, **find the value that falls in the middle of the set.**

If you are using software or a calculator that supports the calculation of the median, you won't need to sort the set of numbers and manually select the middle value.

Here are the same salaries, but this time we'll determine the median:

Rank	Position	Annual Salary
1	Vice President	475,000
2	Manager	165,000
3	Manager	165,000
4	Analyst	65,000
5	Graphic Artist	64,000
6	Analyst	63,000
7	Graphic Artist	62,000
8	Writer	54,000
9	Writer	52,000
10	Admin Assistant	43,000
11	Admin Assistant	39,000
12	Intern	28,000
13	Intern	25,000
	Median Salary	$62,000

FIGURE 2.10 This is an example of the use of the statistical median.

This data set contains 13 values, so the value that resides precisely in the middle is the seventh, which is $62,000. If you want to communicate the typical marketing department salary, $62,000 would do a better job than $100,000. If your purpose is to summarize the salaries of each department in the company to show their comparative impact on expenses, however, which type of average would work better: the median or the mean? In this case the mean would be the better choice because you want a number that fully takes all values into account, including the extremes. To ignore them through use of the median would undervalue the financial impact.

The median is actually an example of a special kind of value called a *percentile*. A percentile expresses the percentage of values that fall below a particular value. The median is another name for the 50th percentile; that is, it expresses the value below which 50% of the values in the set fall.

You may have noticed while considering how to determine the median above that I ignored a potential complication in the process. What do you do if your data set contains an even number of values, rather than an odd number like the 13 employee salaries above? You simply take the *two* values that fall in the middle of the set (e.g., the fifth and sixth values in a set of ten); then, determine the value halfway in between the two. In fact, you can use the same method that you use for calculating the mean to find the value halfway between the two middle values: sum the two middle values then divide the result by two. If you're using software or a calculator to determine the median, this process is handled for you automatically.

MODE AND MIDRANGE

The two remaining types of averages, modes and midranges, are rarely useful in business, but let's take a moment to understand what they are.

The *mode* is simply the value that appears most often in a set of values. In the set of marketing department salaries that we examined previously, the mode is $165,000 because this is the only value that appears more than once in the set. As you can see, the mode wouldn't be a useful means of expressing the middle of marketing department salaries. The most common value in a data set isn't necessarily anywhere near the middle. If no value appears more than once, the set doesn't even have a mode. If two values appear twice in the set, the set is *bimodal*. If more than two values appear more than once with the same degree of frequency, the set is *multimodal*. Modes are rarely useful for business purposes.

The last method for summarizing the middle of a set of values is the simplest to calculate, but you get what you pay for. It's called the *midrange*. The midrange is the value midway between the highest and lowest values in a set of values. To calculate the midrange, you find the highest and lowest values in the set, add them together, and then divide the result by two. This method is an extremely fast way to calculate an average. If you're on the spot for a quick estimate, you can use the midrange. Be careful, though, for unless the values in the set are distributed evenly across the range, the midrange is far too sensitive to the extremes of the highest and lowest values. You're always better off using the mean or the median.

Measures of Distribution

At times you need to communicate more than the center of a set of values. For example, sometimes you need to communicate the degree to which values vary—the range across which the values are distributed. Two sets of values can have exactly the same average value, but one set could be spread across a broad range while the other is tightly grouped around its average. In some cases this difference is significant. Values that vary widely are volatile. Perhaps they shouldn't be, so you're helping the business by pointing it out. For example, if salaries for the same position vary greatly across your company, this may be a problem worth noting and correcting. It may be useful to recognize and communicate to senior management that sales in January for the past 10 years were always only 4% of annual sales, varying no more than half a percent either way from year to year. Such a pattern, with no significant variation, despite expensive marketing campaigns, may indicate that you should save your marketing budget for later in the year. Values that fall far outside the normal range may indicate underlying problems or even extraordinary successes that should be investigated. A salesperson with an unusually high order-return ratio may be selling his customers products they don't need. A department with exceptionally low expenses per employee may have something useful to share with the rest of the company.

The distribution of a set of values can be expressed succinctly through the use of a single number, but there are multiple methods for expressing distributions. We will examine the two that are most useful for business purposes:

- Range
- Standard Deviation

Like averages, these two measures of distribution each work best in specific circumstances. Let's use an example consisting of two sets of values to illustrate these circumstances. Imagine that you work for a manufacturer that uses two

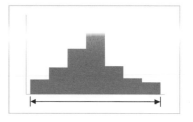

Another term for measures of distribution is *variation*.

warehouses to handle the storage of inventory and the shipping of orders. You've been receiving complaints from customers about the shipments of orders from Warehouse B. To simplify the example, let's say that you've gathered information from each warehouse about shipments of 12 orders of the same product during the same period of time. Ordinarily you would gather shipment information for a much larger number of orders to ensure a statistically significant sample of data, but we'll stick with a small data set to keep the example simple. Here are the relevant values, which in this case are the number of days it took for each of the 12 orders to be processed, from the time each order was received to the time it was shipped:

Order	Days to Ship from	
	Warehouse A	Warehouse B
1	3	1
2	3	1
3	3	1
4	4	3
5	4	3
6	4	4
7	5	5
8	5	5
9	5	5
10	5	6
11	5	7
12	5	10

FIGURE 2.11 This table shows the days it took to ship two sets of 12 orders, one set from Warehouse A and one from Warehouse B.

Because the use of sums and averages is such a common way of analyzing and summarizing quantitative information, you could begin by performing these calculations, resulting in the following:

Warehouse	Sum	Mean	Median
A	51	4.25	4.5
B	51	4.25	4.5

FIGURE 2.12 This table contains various numbers that summarize the number of days it took the two warehouses to each ship a set of 12 orders.

If you were locked into this one way of summarizing and comparing sets of numbers, however, you might conclude and consequently communicate that the service provided by Warehouse B is equal to that of Warehouse A. If you did, you would be wrong.

The significant difference in performance between the two warehouses jumps out at you when you focus on the variation. Warehouse A provides a consistent level of service, always shipping orders in three to five days from the date they're received from the customer. Warehouse B, in contrast, is all over the map. Sometimes it fulfills orders much faster than Warehouse A, and at others times its performance is much slower. It's likely that the complaints came from customers who received their orders after waiting longer than five days and perhaps also from regular customers who, like most, value consistency in service, and find it annoying to receive their orders anywhere from one to 10 days after placing them. Given this message about the inconsistent performance of Warehouse B, let's take a look at the two available ways to measure and communicate this variation.

RANGE

The simpler of the two methods is called the *range*. You can calculate the range as follows:

> Subtract the lowest value from the highest value.

That's it. This is a simple measure of distribution that everyone can understand, which is its strength. To summarize the variation in the performance of Warehouse A versus Warehouse B, you could do so as simply as follows:

	Warehouse A	Warehouse B
Range of days to ship	2	9

FIGURE 2.13 This table shows the ranges of days it took the warehouses to ship the two sets of orders.

Similar to the midrange averaging method, the range method of measuring the distribution of values suffers from its dependence on too few data (only the highest and lowest values), which robs it of the greater accuracy and usefulness of the standard deviation method, which we'll examine next. If Warehouse B had shipped seven orders in five days, including one order in one day and one order in 10 days, that would be a different story from the one contained in our data, but the range would be the same. Nevertheless, everyone can understand a range, which makes it a useful way to communicate distributions to audiences that haven't learned to interpret more complicated measures, such as standard deviations.

STANDARD DEVIATION

The measure of variation that is generally the most useful is the *standard deviation*. Here's a definition:

> The standard deviation of a set of values measures
> their distribution relative to the mean.

The bigger the standard deviation, the greater the range of distribution relative to the mean. This becomes a little clearer when you visualize it. First, take a look at the number of days it took Warehouse B to ship each order compared to the mean value of 4.25 days:

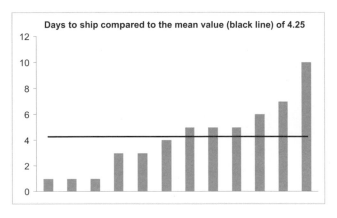

FIGURE 2.14 This graph shows a simple way to visualize the days it took Warehouse B to ship each of the 12 orders compared to the mean value of 4.25 days.

Or, better yet, because our purpose here is to examine the degree to which the shipments of the individual orders varied about the mean, this graph makes it a little easier to visualize:

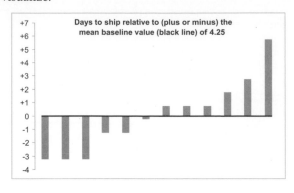

FIGURE 2.15 This graph displays the days it took to ship the individual orders relative to the mean.

So far we haven't displayed the standard deviation. We're still leading up to that. The standard deviation will provide a single value that summarizes the degree to which the 12 shipments as a whole were distributed about the mean (i.e., an average degree of distribution). The standard deviation can be calculated as follows:

1. Calculate the mean of the set of values.
2. Subtract each individual value in the set from the mean, resulting in a list of values that represent the differences of the individual values from the mean.
3. Square each of the values calculated in step 2.
4. Sum the values calculated in the step 3.
5. Divide the value calculated in step 4 by the number of values.
6. Calculate the square root of the value calculated in step 5.[1]

1. These steps were derived from Mario F. Triola (2001) *Elementary Statistics,* Eighth Edition. New York: Addison Wesley Longman Inc.

Technically, there are two formulas for calculating a standard deviation, one for the standard deviation of an entire population of values, and one for the standard deviation of a sample set of values. The steps above are used for an entire population of values. If the value set only includes a sample of the entire population of values, step 5 differs in that you divide by the number of values minus 1, rather than simply by the number of values. It is handy to know how to calculate a value like a standard deviation, but you may never need to do the math yourself. Most software products that produce tables and graphs provide a simple means to calculate the standard deviation.

Because the set of values representing the number of days it took for Warehouse B to ship orders is only a sample set of values (i.e., 12 orders that shipped on a particular day), we'll use the form of the calculation used for sample sets, which produces a standard deviation of 2.58602 days. We can round this figure off to 2.59. This compares to a standard deviation for Warehouse A's shipments of 0.83 days. The difference between a standard deviation of 2.59 and one of 0.83 succinctly indicates a much higher degree of variation in Warehouse B's shipping performance when compared to Warehouse A's. Standard deviations are a concise measure that can be used to compare the relative distribution among multiple sets of values.

In addition to its use for comparisons, a single standard deviation can tell you something about the degree to which the values are distributed. However, to be able to simply look at a standard deviation and interpret the range of variation that it represents requires that you learn a little more about standard deviations.

In general, when individual instances of almost any type of event are measured, and those measurements are arranged by value from lowest to highest, most values tend to fall somewhere near the center (i.e., near some measure of the middle, such as the mean). The farther you get from the center, the fewer the instances you will find. If you display this in the form of a graph called a *frequency polygon*, which uses a line to trace the frequency of instances that occur for each value from lowest to highest, you have something that looks like a *bell-shaped curve*, formally called a *normal distribution* in statistics.

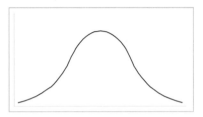

FIGURE 2.16 This curved line represents a *normal distribution*. It displays the frequency of values as they occur from the lowest value at the left to the highest value at the right. Most instances have values near the midpoint of the range of values, which represents the mean. In a perfect normal distribution, the frequency of instances decreases at the same rate to the left and to the right of the mean, resulting a curve (i.e., the black line) that is symmetrical.

The more closely the number of values that you include approaches the entire population of values, the more closely the curve resembles a bell. So what is the significance of a normal distribution to our examination of standard deviations? When you have a normal distribution, the standard deviation describes the distribution of the values as percentages of the whole. The following figure overlays the normal distribution displayed in the figure above with useful information that the standard deviation reveals.

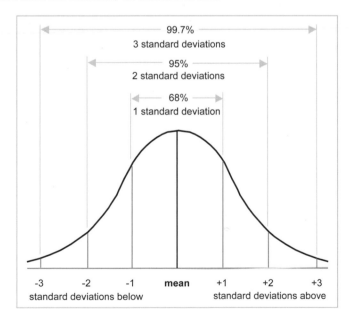

FIGURE 2.17 This figure shows a normal distribution of values in relation to the standard deviations of those values. The percentages of values that fall within one, two, and three standard deviations from the mean can be predicted with a normal distribution, and consequently can be predicted to a fair degree with anything that is close to a normal distribution. This is called the *empirical rule*.

With normal distributions, 68% of the values fall within one standard deviation above and below the mean, 95% fall within two standard deviations, and 99.7% fall within three. Stated differently, if you are dealing with a distribution of values that is close to normal, you automatically know that one standard deviation from the mean represents approximately 68% of the values, two represent approximately 95%, and so on. Given this knowledge, the standard deviation of a set of values has meaning in and of itself, not just as a tool for comparing the degree of distribution between two or more sets of values. The bigger the standard deviation, the broader the range of values, and thus the greater difference in distribution between them.

How does this relate to your world and the types of business phenomena that you study and communicate? Take a couple of minutes to list a few examples that are good candidates for measures of distribution. In what situations does distribution indicate something important to your business?

· · · · · · ·

Here are a few examples that I've encountered:

- Distribution of the *selling price* of specific products or services. Is the distribution greater for some parts of the world or for some sales representatives than for others? Do the differences in distribution correspond to increased or decreased profits?
- Different distributions in measures of *performance*, such as the time it takes to manufacture products, answer phone calls, or resolve technical problems. Do instances of greater distribution indicate problems in training, employee morale, process design, or systems? What does a greater degree of distribution today compared to the past signify?
- Distribution of employee *compensation*. Why is there such a discrepancy in compensation for the same job in different departments? Does the broad distribution of salaries have an effect on employee morale or performance?
- Distribution of the *cost of goods* purchased by various buyers from various vendors. Why is the distribution of costs associated with some buyers so much greater than the distribution associated with others for the same goods?
- Distribution of the departmental *expenses*. How is it that some departments are managing to keep their expenses so much lower than other departments are?

I could go on, but I suspect the point is clear. Measures of distribution tell important stories, so familiarity with the available methods for summarizing and concisely communicating these messages is indeed useful.

Measures of Correlation

Earlier in this chapter, I described *correlation* as a particular type of quantitative relationship where two paired sets of values are compared to one another to see whether they correspond in some manner. For instance, does tenure on the job relate to productivity? In this section we are going to look at a particular way to measure correlation and express it as a single value. This single value is called the *linear correlation coefficient*. It answers each of the following questions about the correlation of two paired sets of quantitative values:

- Does a correlation exist?
- If so, is it strong or weak?
- If so, is it positive or negative?

Here's a concise definition:

> The linear correlation coefficient measures the direction
> (positive or negative) and degree (strong or weak) of the linear
> relationship between two paired sets of values.

By *two paired sets of values* I mean the two sets of values that are involved when you examine the relationship of one thing to another, such as an employee's tenure (e.g., number of years on the job) to his productivity on the job (e.g., number of products manufactured per hour). In this case, the two measures for each employee constitute a paired set of values. By *linear correlation* I mean a consistent relationship between two things; for instance, if you measure the correlation between employee tenure and productivity, and find that as tenure increases productivity increases, or that as tenure increases productivity decreases. However, a linear correlation cannot represent a relationship that varies, for example, if productivity increases along with tenure to a point but after that point declines or levels off. This is clearly a relationship, but it is *nonlinear*. The *direction* of a correlation is either positive or negative. With positive correlations between two sets of values (A and B), as the value of A increases, the value of B likewise increases and as the value A decreases, so does B. With negative correlations, as the value of A increases, the value of B decreases, and vice versa.

If you had to calculate the linear correlation coefficient manually, you would have to work through several steps. Very few of us who work with business numbers need to do so because we have software or calculators to do this for us. What really matters is that we know how to interpret the resulting value, so let's focus on the number itself and what it means.

Despite its intimidating name, the linear correlation coefficient is actually quite simple to interpret. Here are a few guidelines:

- All linear correlation coefficient values fall somewhere between +1 and -1.
- A value of 0 indicates that there is no correlation.
- A value of +1 indicates that there is a perfect positive correlation.
- A value of -1 indicates that there is a perfect negative correlation.
- The greater the value, in either the positive or the negative direction, the stronger the correlation.

Pretty simple, but it will still help to look at this visually. To do so, we're going to use a graph called a *scatter plot*, which is designed specifically to display the correlation of two paired sets of quantitative values. Perhaps you've seen this type of graph listed as one that is available in software that you use but have never used it, and perhaps have only a vague idea how it works. With a little exposure, you'll find that scatter plots are quite easy to use and interpret as well as quite useful for revealing and communicating quantitative relationships.

Here's a series of scatter plots that will help you visualize the types of relationships that a linear correlation coefficient is designed to reveal. Each graph displays the relationship between two paired sets of values, one horizontally along the X axis and one vertically along the Y axis. When you read a scatter plot, you should look for what happens to the value along the X axis in relation to the value along the Y axis. As X goes up, what happens to Y? As X goes down, what happens to Y? Is the relationship strong (i.e., close to a straight line) or is it weak (i.e., bounces around)? Is it positive (i.e., moves upward from left to right) or is it negative (i.e., moves downward from left to right)? Each of the following graphs displays a different relationship between the variable plotted along the X axis (horizontal) and the variable plotted along the Y axis (vertical), with the linear correlation coefficient in parentheses to help you understand its meaning.

A *variable* is simply something with values that can vary, such as employee productivity.

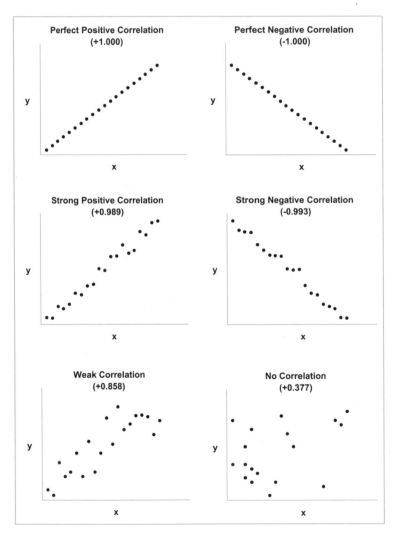

FIGURE 2.18 This is a series of scatter plots, each of which displays a different relationship between two sets of paired values (e.g., employee tenure and productivity).

Bear in mind that these scatter plots simply provide examples of correlations. If the linear correlation coefficient in the left-middle scatter plot was +0.970 rather than +0.989, it would still represent a strong positive relationship.

One way of looking at correlations as displayed in scatter plots is to imagine a straight line that passes through the center of the dots; then, determine the strength of the correlation based on the degree to which the dots are tightly grouped around that line: the tighter the grouping, the stronger the relationship. Here are examples of how scatter plots would look if you actually drew the lines:

Drawing a straight line of *best fit* through the center of a series of points on a scatter plot is a common technique for highlighting the relationships between two sets of values. It's called a *trend line, line of best fit,* or, more formally, a *regression line*.

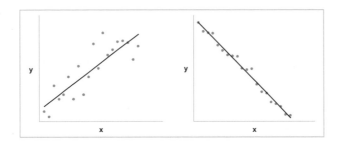

FIGURE 2.19 These are scatter plots with lines of best fit through the center of the dots to clearly delineate the nature of the relationship.

Based on what you've learned about scatter plots, how would you describe each of the relationships displayed above?

. . . .

In the scatter plot on the left, the characteristics that you must consider are:

- The direction of the line, which in this case is upward from left to right
- The closeness of the grouping of dots around the line, which in this case is not terribly tight

Given these two observations, we can say that the scatter plot on the left depicts a correlation that is positive (i.e., upward from left to right) but moderately strong (i.e., not tightly grouped around the line). Using this same method of interpretation, the scatter plot on the right depicts a correlation that is negative and very strong but not perfectly so.

At this point, you may be wondering: "At what value of a linear correlation coefficient does a correlation cease to be strong and begin to become weak or cease to be weak or even a correlation at all?" There is no precise answer to this question. It depends to some degree on the number of paired values included in your data sets; the more values you have, the greater confidence you can have in the validity of the linear correlation coefficient. Because our purpose here is not to delve too deeply into the realm of statistics, let's be content with the knowledge that values close to 1 in positive correlations and close to -1 in negative correlations indicate strong relationships and that the closer they are to 1 or -1, the stronger the relationship.

Remember, linear correlation coefficients can only describe relationships that are linear—that is, ones that move in one direction or another—but not relationships that are positive under some circumstances and negative under others. Here's such an example:

For an excellent introduction to statistics, including much more information than I've provided about correlations, I recommend the textbook by Mario F. Triola (2001) *Elementary Statistics*, Eighth Edition. New York: Addison Wesley Longman Inc.

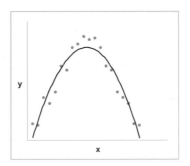

FIGURE 2.20 This is an example of a nonlinear correlation.

What you see here is definitely a correlation, but it certainly isn't linear. If this scatter plot represents the relationship between employee tenure (i.e., years on the job) on the X axis and employee productivity on the Y axis, how would you interpret this relationship, and how might you explain what is happening to productivity after employees reach a certain point in their tenure?

. . . .

After studying this scatter plot and double-checking the data, you would likely suggest that something be done as employees reach the halfway point along their tenure timelines, such as offering new incentives to keep them motivated or retraining for new positions that they might find more interesting.

Measures of Ratio

In contrast to correlations, which measure the relationship between multiple paired sets of values, a ratio measures the relationship between a single pair of values. A typical example that we encounter in business is the book-to-bill rate, which is a comparison between the value associated with sales orders that have been booked (i.e., placed by the customer and accepted as viable orders) and the value associated with actual billings that have been generated in response to orders.

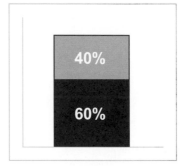

Ratios can be expressed in four ways:

- As a *sentence*, such as "Two out of every five customers who access our web site place an order."
- As a *fraction*, such as 2/5 (i.e., 2 divided by 5)
- As a *rate*, such as 0.4 (i.e., the result of the division expressed by the fraction above)
- As a *percentage*, such as 40% (i.e., the rate above multiplied by 100, followed by a percent sign)

Each of these expressions is useful in different contexts, but rates and percentages are the most concise and therefore the most useful for tables and graphs. Many measures of ratio have conventional forms of expression, such as the book-to-bill rate mentioned above, which is typically expressed as a rate (e.g., 1.25, which indicates that for every five orders that have booked, only four have been billed, or 5/4 = 1.25), or the profit margin, which is normally expressed as a percentage (e.g., 25%, which indicates that for every $100 of revenue, $75 goes toward expenses, leaving a profit of $25, or $25/$100 = 0.25 * 100 = 25%).

Take a moment to think about and list a few of the ways that ratios are used, or could be used, to communicate quantitative information related to your own work.

. . . .

Ratios are simple shorthand for expressing the direct relationship between two values. One especially handy use of ratios is to compare several individual values to a particular value to show how they differ. In this case, your purpose is not to compare the actual values but to show the degree to which they differ. In such circumstances, you can simplify the message by setting the main value to which you are comparing all the others to a baseline of 1 (expressed as a rate) or 100% (expressed as a percentage); then, express the other values as ratios that fall above or below that baseline. Here's an example expressed in percentages:

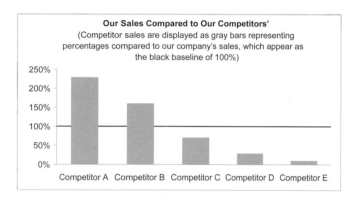

FIGURE 2.21 This graph includes a baseline of 100% for the primary set of values, making it easy to see how the other values, also expressed as percentages, differ.

By using a baseline, it is easy to see that the main competitor does about 250% of your company's value, which is 2.5 times as much when expressed as a rate. Expressing the comparison in this manner eliminates the need for readers to do calculations in their heads when they want to think in terms of relative differences.

Measures of Money

Most quantitative information that we encounter in business involves some currency of exchange—in other words, money. Be it U.S. or Canadian dollars, Japanese yen, British pounds, Swiss francs, or the newer Euro, money is at the center of most business analysis and reporting. Unlike most other units of measure, currency has a characteristic that we must keep in mind when communicating information that spans time: the value of money is not static; it changes with time. The value of a U.S. dollar in November of 2001 was not the same as its value in November of 2002. If you've been asked to prepare a report that exhibits the trend of sales in U.S. dollars for the past three years, would you be justified in asserting that sales have increased by 20% during that time if three years ago annual sales were $100 million and today they total $120 million? Only if the value of a dollar today is the same as it was three years ago, which it isn't.

When the value of a dollar decreases over time, we refer to this as *inflation*. When comparing money across time, an accurate comparison can only be made when you adjust for inflation. I've found, however, that in business reporting this is rarely done. Despite the validity of the argument in favor for adjusting for inflation, doing so isn't always practical, so I won't attempt to force on you a practice that you may very well ignore. For those of you who can take extra time required to correct for skewed results due to inflation, I've included Appendix C,

Adjusting for Inflation, in the back of the book. It isn't difficult, and the practice will improve the quality of your financial reporting.

Business today, especially in large companies, is often international and entails multiple currencies. This is a problem when we must produce reports that combine data across multiple currencies, such as sales in the Americas, Europe, and Asia. You can't just throw the numbers together because 100,000 U.S. dollars does not equal 100,000 British pounds or 100,000 Japanese yen. To combine them or to compare them, you must convert them all into a single currency. Fortunately, most operational software systems that we use to run our businesses today are designed to do this work for us, converting money based on tables of exchange rates, so we can easily see transactions both in their original currency and in some common currency used for international reporting, such as U.S. dollars. Because software typically does this work for us, my intention here is simply to caution you to avoid mixing currencies without converting them to a common currency. If you're not careful, you could inadvertently report results that are in error by a large order of magnitude.

When your purpose is to compare monetary values, such as those associated with different categories (geographical area, departments, etc.), you can often avoid the challenge of mixing multiple currencies by expressing the numbers in the form of rates or percentages. For instance, if you want to compare the annual sales through your various sales channels (direct sales force, distributors, etc.) for the past three years by expressing the sales of each sales channel as a percentage of the whole rather than as currency, you not only avoid all problems associated with inflation and multiple currencies, you also present the numbers in a manner that speaks the message of comparative sales directly and clearly. If you wanted to see such a comparison, would you prefer this table . . .

Channel	2001	2002	2003
Direct	388,838	476,303	593,838
Reseller	546,373	501,393	504,993
OEM	85,303	99,383	150,383
Total	$1,020,514	$1,077,079	$1,249,214

FIGURE 2.22 This table displays a comparison of sales by sales channel, expressed as dollars.

. . . or this one?

Channel	2001	2002	2003
Direct	38%	44%	48%
Reseller	54%	47%	40%
OEM	8%	9%	12%
Total	100%	100%	100%

FIGURE 2.23 This table displays a comparison of sales by sales channel, expressed as percentages of total sales.

As you can see, the second approach completely eliminated the need to account for inflation and multiple currencies.

Understanding the relationships we've examined in this chapter lays the foundation that will help you design tables and graphs to effectively communicate quantitative information. In the next chapter, we'll look at the basics of tables and graphs and begin to see how they can effectively present the kinds of relationships we've just discussed in Chapter 2.

Summary at a Glance

Quantitative Relationships

- Quantitative information consists of two types of data:
  - Quantitative
  - Categorical
- Quantitative information always describes relationships.
- These relationships involve either
  - Simple associations between quantitative values and categorical subdivisions or
  - More complex associations among multiple sets of quantitative values.
- There are four types of relationships within categories:
  - Nominal
  - Ordinal
  - Interval
  - Hierarchical
- There are three types of relationships between quantitative values:
  - Ranking
  - Ratio
  - Correlation

Numbers that Summarize

Type of Summary	Method	Note
Average	Mean	Measures the center of a set of values in a manner that is equally sensitive to all values, including extremes
	Median	Measures the center of a set of values in a manner that is insensitive to extreme values
Distribution	Range	Simple to calculate, relying entirely on the highest and lowest values, but only roughly defines a ranges of values
	Standard Deviation	Provides a rich expression of the distribution of a set of values across its entire range
Correlation	Linear Correlation Coefficient	Indicates whether a correlation exists between two paired sets of values, and if so, its direction (positive or negative) and its strength (strong or weak)
Ratio	Rate or Percentage	Measures the direct relationship between two quantitative values

Measures of Money

- When comparisons of monetary value are expressed across time, adjusting the value to account for inflation produces the most accurate results.
- When reporting monetary values that combine multiple currencies, you must first convert them all into a common currency.

3 FUNDAMENTAL CONCEPTS OF TABLES AND GRAPHS

Tables and graphs are the two basic forms for communicating quantitative information. They have developed over time to the point where we now thoroughly understand which works best for what type of information and why. This chapter introduces tables and graphs and gives simple guidelines for selecting one form over the other.

Quantities and categories
Tables defined
When to use tables
Graphs defined
A brief history of graphs
When to use graphs

Tables and graphs are the two primary means to structure and communicate quantitative information. Both formats have been around for quite some time and have been researched extensively to hone their use to a fine edge of effectiveness. The best practices that have emerged are not difficult to learn, understand, and put to use in your everyday work with numbers.

Occasionally, the best way to display quantitative business information is not in the form of a table or graph. When the quantitative information you want to convey consists only of a single number or two, written language is an effective means of communication; your message can be expressed simply as a sentence or highlighted as a bullet point. If your message is that last quarter's sales totaled $1,485,393 and exceeded the forecast by 16%, then it isn't necessary to structure the message as a table, and there is certainly no need to create a graph. You can simply say something like:

Q2 sales = $1,485,393, exceeding forecast by 16%

Alternatively, it wouldn't hurt to structure this message in simple tabular form such as this:

Q2 Sales	Compared to Forecast
$1,485,393	16%

FIGURE 3.1 This table shows sales values, arranged in columns.

or this:

Q2 Sales	$1,485,393
Compared to Forecast	+16%

FIGURE 3.2 This table shows sales values, arranged in rows.

If you're like a lot of folks, however, you might be tempted while structuring this information as a table to jazz it up in a way that actually distracts from your simple, clear message—perhaps something like this:

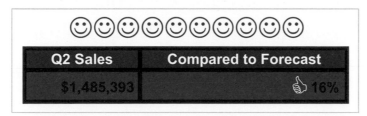

FIGURE 3.3 This table shows sales values, arranged to impress, or perhaps to entertain, but certainly not to communicate.

You might even be tempted to pad the report with an inch-thick stack of pages containing the details of every sales order received during the quarter, eager to demonstrate how hard you worked to produce those two sales numbers. However, as we'll observe many times in this book, there is *elegance in simplicity*.

Quantities and Categories

Before we launch into an individual examination of tables and graphs, let's touch on an important feature that is common to both. Quantitative messages are made up of two types of data:

- Quantitative
- Categorical

Quantitative values are numbers, measures of something related to the business (number of orders, amount of profit, rating of customer satisfaction, etc.). Categories identify what the quantitative values measure. These two types of data fulfill different roles in tables and graphs.

In the following example of a simple table that displays employee compensation by department, split between exempt and non-exempt employees, all the data are either quantities being measured or subdivisions of categories to which the quantities relate. Can you identify which is which?

Department	Exempt	Non-Exempt
Sales	950,003	1,309,846
Operations	648,763	2,039,927
Manufacturing	568,543	2,367,303
Total	$2,167,309	$5,717,076

FIGURE 3.4 This is a table of employee compensation information that you can use to practice distinguishing quantitative and categorical data.

The labels that identify the various departments in the leftmost column, including *Total*, and *Exempt* and *Non-exempt* are categorical subdivisions. They establish the context for the numbers. The term *Department* in the upper left corner names the category to which the categorical subdivisions *Sales*, *Operations*, and *Manufacturing* belong. All other data (i.e., all the numbers in this table) are quantitative.

Do numbers always represent quantitative values? No, not always. Sometimes numbers simply categorize information and have no quantitative meaning. Order numbers (e.g., 1003789), numbers that identify the year (e.g., 2003), and numbers that sequence information (number 1, number 2, number 3, etc.), to name a few examples, express categorical data. One simple test to determine this distinction is to ask the question, "Would it make sense to add these numbers up, or to perform any other type of math on them?" For instance, would it make sense to add up order numbers? No, it wouldn't. How about numbers that rank a list of sales people as number 1, number 2, etc.? Once again, the answer is no; therefore, these numbers represent categorical, rather than quantitative, data.

Let's take a look at one more example to practice making the distinction between categorical and quantitative information. Using the graph below, test your skills by identifying which components represent categorical subdivisions, and which represent quantitative values.

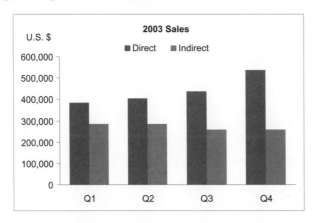

FIGURE 3.5 This graph depicts sales information for practice in distinguishing quantitative and categorical data.

This exercise is a little more difficult than the previous one. Let's take it one component at a time. For each of the items below, indicate whether the information is of the categorical or quantitative type.

1.   The values of time along the bottom (*Q1*, *Q2*, etc.)
2.   The dollars along the left side
3.   The legend, which encodes *Direct* and *Indirect*
4.   The vertical bars in the body of the graph
5.   The title and subtitle in the top center

Once you've identified each, take a moment to compare your answers to those on the right.

· · · · · · ·

Did you catch the dual role of the bars, that they contain both quantitative values and categorical subdivisions? With a little practice, you will be able to easily deconstruct graphs into quantitative and categorical data. This ability will enable you to apply the differing design practices that pertain to each type of data.

Tables Defined

A table is a structure for organizing and displaying information; a table exhibits the following characteristics:

ANSWERS
1. Categorical, labeling the quarters of the year
2. Quantitative, providing dollar values for the interpretation of the heights of the bars
3. Categorical, providing a distinction between direct and indirect sales
4. Both quantitative and categorical; the heights of the bars encode quantitative information about sales in dollars; the colors of the bars encode categorical data identifying which sales are direct vs. indirect
5. Categorical, identifying the year of the sales

- Data are arranged in columns and rows.
- Data are encoded as text (including words and numbers).

A single set of values, occupying a single column or row, is simply a list, not a table.

Although the column and row structure of tables is often visually reinforced by grid lines (i.e., horizontal and vertical lines outlining the columns and rows), it is the arrangement of the information that characterizes tables, not the presence of lines that visibly delineate the structure of the underlying grid. In fact, as we will see later in the chapter on table design, grid lines must be used with care to keep them from diminishing a table's usefulness.

Tables are not used exclusively to display quantitative information. Anytime you have more than one set of values, and a relationship exists between values in the separate sets, you may use a table to align the related values by placing them in the same row or column. For instance, tables are often used to display meeting agendas, with start times in one column, the name of the topic that will be covered in the next, and the name of the facilitator in the next, such as the following:

Time	Topic	Facilitator
09:00 AM	Opening remarks	Scott Wiley
09:15 AM	Product demo	Olivia Presosil
10:00 AM	Discussion	Jerry Snyder
10:45 AM	Planning	Pamela Smart

FIGURE 3.6 This is an example of a table that does not contain quantitative data. In this case a meeting agenda.

Tables have been in use for centuries, so they are readily understood by almost everyone who can read.

When to Use Tables

A handful of conditions should direct you to select a table, rather than a graph, as the appropriate display structure, but tables offer one primary benefit:

> Tables make it easy to *look up* values.

Tables excel as a means of displaying *simple relationships between quantitative values and the categorical subdivisions to which these values are related* so that the values can be individually located and considered.

When deciding what structure to employ to communicate your quantitative message, you should always ask yourself how the information will be used. Will you or others use it to look up one or more particular values, or might it be used to examine the quantitative values more holistically to discern patterns, which would make it a prime candidate for display in a graph, as we'll see soon?

Tables also make it easy to compare pairs of related values (e.g., sales in quarter 1 and sales in quarter 2). Here's a typical example:

Lender	Fixed Rate	Adjustable Rate
Bank A	7.500%	6.250%
Bank B	7.375%	6.125%
Bank C	8.375%	7.500%

FIGURE 3.7 This is an example of a simple table used to look up loan interest rates of multiple lenders.

Tables work well for look-up and local comparisons, in part because their structure is so simple, but also because the quantitative values are encoded as text, which we are able to understand directly, without the need for translation. Graphs, by contrast, are visually encoded, which requires some translation of the information into the numbers it represents.

The textual encoding of tables also offers a level of *precision* that cannot be provided by graphs. It is easy to express a number with as much specificity as you wish using text (e.g., 27.387483), but the visual encoding of individual numbers in graphs doesn't lend itself to such precision.

Another strength of tables is that they can include multiple sets of quantitative values that are expressed in *different units of measure*. For instance, if you need to provide sales information that includes the number of units sold, the dollar amount, and a comparison to a forecast expressed as a percentage, doing so in a single graph would be difficult, because a graph usually contains a single quantitative scale with a single unit of measure.

To summarize, tables are used to display simple relationships between quantitative values and corresponding categorical subdivisions, which makes tables ideal for looking up and comparing individual values. The entire set of reasons to use a table consists of the following; if any one of these is true, you should be inclined to choose a table as your means of display.

1. The document you produce will be used to look up individual values.
2. It will be used to compare individual values.
3. Precise values are required.
4. The quantitative information to be communicated involves more than one unit of measure.

Graphs Defined

A graph is a method for displaying quantitative information that exhibits the following characteristics:

- Values are displayed within an area delineated by one or more axes.
- Values are encoded as visual objects positioned in relation to the axes.
- Axes provide scales (quantitative and categorical) that are used to assign values and labels to the visual objects.

Axes delineate the space that is used to display data in a graph.

The defining essence of a graph is that it is a *visual display* of quantitative information. Whereas tables encode quantitative values as text, graphs encode quantitative values visually. Let's look at a simple example:

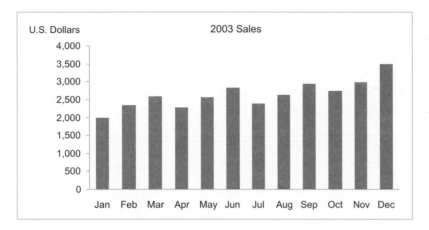

FIGURE 3.8 This is an example of a simple graph, which displays sales information for the year 2003, measured in U.S. dollars.

This graph has two axes: one that runs horizontally across the bottom, called the *X axis*, and one that runs vertically up the left side, called the *Y axis*. In this graph, the categorical scale, which identifies the months, resides on the X axis, and the scale for the quantitative values (i.e., sales in U.S. dollars) resides on the Y axis. The values themselves are encoded as vertical bars. Bars are only one visual object that can be used to encode data in graphs. The quantitative value of each bar is encoded as its length, with the top end of the bar located in a position corresponding to its numeric value on the Y axis. The horizontal position of each bar along the X axis is labeled to denote the specific categorical subdivision to which the sales values are related (e.g., *Jan*).

With a little practice, even someone who has never previously used a graph can learn to interpret the information contained in a simple one like this. Although the values of sales for each month cannot be interpreted to the exact dollar, this isn't the graph's purpose. Rather, the graph paints the clear picture that sales have generally increased during the course of the year, with slight decreases in the first month of each new quarter, consistent increases in the next two months of each quarter, and consistent increases from one quarter to the next, resulting in the best sales during the last quarter. Information like this, which is intended to show patterns in the data, is best presented in a graph rather than a table, as we'll see in the section *When to Use Graphs* below. But first, a little historical background about graphs.

A Brief History of Graphs

Graphs of quantitative information have been in use for a few hundred years, which is a relatively short time given the thousands of years that mathematics has existed. Despite how natural it may seem to us to see quantitative information displayed in graphs, the original notion that numbers could be displayed visually in relation to two perpendicular axes involved a leap of imagination. The launching pad for this leap had already been around for many centuries before quantitative graphs emerged. Another type of visual information display, also assisted by a scale of measurement along perpendicular axes, eventually suggested the possibility of graphs. Can you guess what it was? It is still in common use today for the purpose of navigation. It is a two-dimensional representation of the physical world and can be used to measure distances between locations. I'm referring to a *map*. The earliest known map dates back about 4,300 years. It was drawn on a clay tablet and represented northern Mesopotamia. When a map depicts the entire world, or some large part of it, the standard set of grid lines that allow us to determine location and distance are longitudes and latitudes. These grid lines form a quantitative scale of sorts.

It wasn't until the 17th century that two-dimensional visual grids were first used purely to represent numbers. They were introduced by Rene Descartes, in his *La Geometrie* (1637), as a means to visually encode numbers as grid coordinates. His innovation provided the groundwork for an entire new field of mathematics that is based on graphs.

It was William Playfair, a British social scientist, who used his imagination and design acumen in the late 18th century to invent many of the graphing techniques that we use today. Playfair pioneered the use of graphs to display the shape of quantitative information, thus providing a way to communicate quantitative relationships that numbers in the form of text could not. The old saying, "A picture is worth a thousand words" applies quite literally to graphs. By presenting quantitative information in visual form, graphs efficiently communicate what might otherwise require a thousand words or even a million words—and sometimes what no number of words could ever convey.

From the time of Playfair until today, many innovators have added to the inventory of graphic designs available for the representation, exploration, and communication of quantitative information. In recent years, though, none has contributed more to the field as an advocate of excellence in graphic design than Edward Tufte, who in 1983 published his landmark treatise on the subject, *The Visual Display of Quantitative Information*. With the publication of additional books and many articles since, Tufte continues to be respected by many as the top authority in this field.

Besides the work of Tufte, the work of William S. Cleveland, especially his book *The Elements of Graphing Data*, is also an outstanding resource, especially for those who seek an advanced understanding of graphs and their use.[1] Cleveland's work is especially useful to those interested in sophisticated graphs that are used in scientific research.

Information about the early development of the graph, including its precursor, the map, may be found in Robert E. Horn (1998) *Visual Language*. Bainbridge Island WA: MacroVU, Inc. Robert Horn provides an informative timeline, which cites the milestones in the historical development of visual information display.

1. William S. Cleveland (1994) *The Elements of Graphing Data*. Summit NJ: Hobart Press.

When to Use Graphs

Graphs display quantitative information in a manner that reveals much more than a collection of individual values. Because of their visual nature, graphs present the overall *shape* of the data. Text, as in tables, cannot present the shape of information.

The patterns revealed by graphs enable readers to detect many points of interest in a single collection of information. Take a look at the next example and try to identify some of the features that graphs can reveal. Approach this by

first determining the messages in the data, then by identifying the visual cues—
the shapes—that revealed each of these messages.

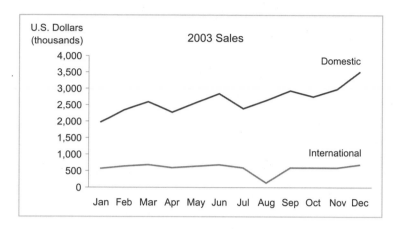

FIGURE 3.9 This fairly typical line graph is an example of the shapes and patterns in quantitative data that graphs make visible.

What insights are brought to your attention by the shape of this information?
Take a moment to list them in the right margin.

. . . .

Let's walk through a few of these revelations together, beginning exclusively
with domestic sales. What does the shape of the black line tell you about domestic sales during the year 2003? One message is that during the course of the year
the general trend of domestic sales was upwards; these sales ended higher than
they started, with a gradual increase through most of the year. One name for this
type of pattern is a *trend*, which displays *change* over time.

Was this upward trend completely steady? No, it exhibits a pattern that
salespeople sometimes call the *hockey stick*. The trend goes down in the first
month of each quarter and then gradually goes up to its quarterly peak in the
last month of each quarter. If you examine the shape of the line from the last
month of one quarter, such as *March*, to the last month of the next quarter, such
as *June*, you'll recognize that it looks a little like a hockey stick. This trend occurs
quarterly, at a finer level of detail than the overall annual trend upward.

Now, if you look at sales in the first quarter of the year versus sales in the
second quarter, then the third, and finally the last, you find that the graph
makes it easy to see how they compare. This is another relationship that the
shape of data brings to light: *comparison*. In this case the comparison is between
different subsets (i.e., quarterly domestic sales) of the same overall set of values
(i.e., domestic sales).

Now look for a moment at international sales. Once again you are able to
easily see the trend of sales through the year, which in this case is relatively
steady. Compared to domestic sales, international sales show much less fluctua-
tion through time and relatively little difference between the beginning and end
of the year. International sales seem to be much less affected by seasons in
general, but one point on the line stands out as quite different from the rest: the
month of August. Sales in August took an uncharacteristic dip compared to the

rest of the year. As an analyst, this point of exception would make you want to dig for the cause. If you did, perhaps you would discover that most of your international customers were vacationing in August and weren't around to place orders. Whatever the cause, the current point of interest to us is that graphs make *exceptions* to general patterns stand out clearly from the rest. This wouldn't be nearly so obvious in a table of numbers.

Finally, if you widen your perspective to all the quantitative information, you find that the graph makes it easy to see the *similarities* and *differences* between the two sets of values (domestic and international sales), both overall and at particular points in the graph. We could go on, adding more terms to the list of characteristics that graphs reveal, but we've hit the high points and are ready to distill what we've detected to its essence.

> Graphs are used to display relationships among multiple
> quantitative values by giving them shape.

The visual nature of graphs endows them with their unique power to reveal patterns of various types, including changes, differences, similarities, and exceptions. Graphs can communicate quantitative relationships that are much more complex than the simple association between individual quantitative values and categorical subdivisions that are expressed in tables.

Graphs can display large data sets in a way that can be readily perceived and understood. You could gather data regarding the relationship between employee productivity and the use of two competing software packages, involving millions of measures across several years, and with the help of a graph, you would be able to immediately see the nature of the relationship. If you have ever tried to use a huge table of data for analysis, you would quickly fall in love with graphs like scatter plots, which can make the relationships among millions of individual data points instantly intelligible.

Summary at a Glance

Use Tables When	Use Graphs When
• The document will be used to look up individual values.	• The message is contained in the shape of the values.
• The document will be used to compare individual values.	• The document will be used to reveal relationships among multiple values.
• Precise values are required.	
• The quantitative information to be communicated involves more than one unit of measure.	

4 FUNDAMENTAL VARIATIONS OF TABLES

Tables are structured according to the nature of the information they are meant to display. This chapter breaks tables down into their fundamental variations and gives simple rules of thumb for pairing your message with the best tabular means to communicate it.

Relationships in tables
   Quantitative-to-categorical relationships
   Quantitative-to-quantitative relationships
Variations in table design
   Unidirectional
   Bidirectional
   Table design solutions

In this chapter we will first take a look at *what* tables can be used to display, followed by *how* tables can be structured visually; then, we'll link the *what* to the *how*.

Relationships in Tables

Information that we display in tables always exhibits a simple relationship between individual values. Tables are commonly used to display five relationships, which can be divided into two basic types:

- Quantitative-to-categorical relationships
  - Between one set of quantitative values and one set of categorical subdivisions
  - Between one set of quantitative values and the intersection of multiple categories
  - Between one set of quantitative values and the intersection of hierarchical categories
- Quantitative-to-quantitative relationships
  - Among one set of quantitative values associated with multiple categorical subdivisions
  - Among distinct sets of quantitative values associated with the same categorical subdivision

This no doubt sounds like gobbledygook at the moment, but the meaning of each of these relationships should become clear in the sections below.

Quantitative-to-Categorical Relationships

These relationships involve looking up one quantitative value at a time; each value relates either to a single category or to the intersection of multiple categories.

BETWEEN ONE SET OF QUANTITATIVE VALUES AND ONE SET OF CATEGORICAL SUBDIVISIONS

The simplest type of relationship found in tables is an association between quantitative values and a single set of categorical items. Typical examples include sales dollars (quantitative values) associated with geographical sales regions (categorical items), ratings of performance (quantitative values) associated with individual employees (categorical items), or expenses (quantitative values) associated with fiscal periods (categorical items). Here's an example:

Salesperson	QTD Sales
Robert Jones	13,803
Mandy Rodriguez	20,374
Terri Moore	28,520
John Donnelly	34,786
Jennifer Taylor	36,973
Total	$134,456

FIGURE 4.1 This table displays a simple relationship between quantitative and categorical values, in this case sales in dollars per salesperson.

Tables of this type function as *simple lists* used for look-up. You want to know how much Mandy has sold so far this quarter? You simply find Mandy's name on the list, and then look at the sales number next to her name. Tables of this sort are a common means of breaking quantitative information down into small chunks so we can see the details behind the bigger picture.

BETWEEN ONE SET OF QUANTITATIVE VALUES AND THE INTERSECTION OF MULTIPLE CATEGORIES

One small step up the complexity scale, tables can also display quantitative values that are simultaneously associated with multiple sets of categorical subdivisions. Here's the same example as the one above, but this time a second category has been added:

Salesperson	Jan	Feb	Mar
Robert Jones	2,834	4,838	6,131
Mandy Rodriguez	5,890	6,482	8,002
Terri Moore	7,398	9,374	11,748
John Donnelly	9,375	12,387	13,024
Jennifer Taylor	10,393	12,383	14,197
Total	$35,890	$45,464	$53,102

FIGURE 4.2 This table displays a relationship between quantitative values and multiple categories, in this case between sales dollars and particular salespersons in a particular month.

In this example, each quantitative value measures the sales for a particular salesperson in a particular month. A single quantitative value measures sales for the intersection of two categories. This is still a relationship between quantitative values and categorical subdivisions, but it is slightly more complex in that it addresses the intersection of multiple categories. Because each category represents a different dimension of the business, this type of table provides a multi-dimensional view of the business.

BETWEEN ONE SET OF QUANTITATIVE VALUES AND THE INTERSECTION OF HIERARCHICAL CATEGORIES

A variation of the relationship in the previous example occurs when multiple sets of categorical subdivisions relate to one another hierarchically. For instance, let's consider the following three categories: years, quarters, and months. They are related hierarchically in that years consist of quarters, which, in turn, consist of months. Another common example involves the hierarchical categories that organize and describe products. These may consist of product lines, which, in turn, consist of product families, which, in turn, consist of individual products. Just as in the previous type of relationship, a quantitative value measures the intersection of categories. The only difference is that in this case there is a hierarchical relationship among sets of categorical subdivisions. Here's an example:

Product Line	Product Family	Product	Value
Hardware	Printer	PPS	6,131
		PXT	8,002
		PQT	11,748
	Router	RRZ	13,024
		RTS	14,197
		RQZ	23,293
Software	Business	ACT	12,393
		SPR	9,393
		DBM	5,392
	Game	ZAP	10,363
		ZAM	15,709
		ZOW	13,881
Total			$143,526

FIGURE 4.3 This table displays a relationship between quantitative values and hierarchical categories, in this case between sales dollars and a hierarchically arranged combination of product line, product family, and product.

Quantitative-to-Quantitative Relationships

In addition to permitting us to show direct relationships between quantitative values and categorical subdivisions, tables also allow us to examine relationships among multiple quantitative values. These relationships involve the comparison of multiple quantitative values that measure either the same entity (e.g., expenses in U.S. dollars) for multiple categorical subdivisions (e.g., the months January and February), or multiple entities (e.g., expenses and revenues in U.S. dollars) for a single categorical subdivision (e.g., the month of January).

AMONG ONE SET OF QUANTITATIVE VALUES ASSOCIATED WITH MULTIPLE CATEGORICAL SUBDIVISIONS

Rather than merely looking at the sales associated with an individual salesperson, you may want to compare sales among different salespersons. In the next example, how did Mandy do compared to Terri in February?

Salesperson	Jan	Feb	Mar
Robert Jones	2,834	4,838	6,131
Mandy Rodriguez	5,890	6,482	8,002
Terri Moore	7,398	9,374	11,748
John Donnelly	9,375	12,387	13,024
Jennifer Taylor	10,393	12,383	14,197
Total	$35,890	$45,464	$53,102

FIGURE 4.4 This table examines a relationship among quantitative values associated with multiple categorical subdivisions, in this case sales by several salespeople in the months of January, February, and March.

As you can see, the focus here is on comparisons.

AMONG DISTINCT SETS OF QUANTITATIVE VALUES ASSOCIATED WITH THE SAME CATEGORICAL SUBDIVISION

Tables often contain more than one distinct set of quantitative values. When this is the case, we are able to examine the relationships among these different quantitative measures within the same categorical subdivision. The example below now includes three sets of quantitative values: sales, returns, and net sales:

Salesperson	Sales	Returns	Net Sales
Robert Jones	13,803	593	13,210
Mandy Rodriguez	20,374	1,203	19,171
Terri Moore	28,520	10,393	18,127
John Donnelly	34,786	483	34,303
Jennifer Taylor	36,973	0	36,973
Total	$134,456	$12,672	$121,784

FIGURE 4.5 This table examines a relationship among distinct quantitative values associated with a single categorical subdivision, in this case the salespeople's sales and returns as well as the resulting net sales.

A tabular display like this makes it easy to see how each salesperson is doing in terms of sales versus returns and the resulting net sales.

———————

Here's a summary of what we've learned about *what* tables can be used to display:

Function	*Relationship Type*	*Relationship*
Look-up	Quantitative-to-Categorical	Between a single set of quantitative values and a single set of categorical subdivisions
		Between a single set of quantitative values and the intersection of multiple categories
		Between a single set of quantitative values and the intersection of hierarchical categories
Comparison	Quantitative-to-Quantitative	Among a single set of quantitative values associated with multiple categorical subdivisions
		Among distinct sets of quantitative values associated with the same categorical subdivision

It is important to understand these different types of relationships because, as we'll see in the next section, the layout of a table is affected by the types of relationships the table is intended to show.

Variations in Table Design

Now let's look at *how* tables can display the types of relationships we've identified. Within the general structure of columns and rows, tables can vary somewhat in design. These structural variations can be grouped into two fundamental types:

- Unidirectional—categorical items are laid out in one direction only (i.e., either across columns or down rows)
- Bidirectional—categorical items are laid out in both directions

Unidirectional

When a table is structured unidirectionally, categorical subdivisions are arranged across the columns or down the rows but not in both directions. Unidirectional tables may have one or more sets of categorical subdivisions, but if there is more than one set, the subdivisions are still only arranged across the columns or down the rows. Here is an example of a unidirectional table with categorical subdivisions (i.e., departments) arranged down the rows:

Department	Headcount	Expenses
Finance	26	202,202
Sales	93	983,393
Operations	107	933,393
Total	234	$2,118,988

FIGURE 4.6 This table is structured unidirectionally with the categorical subdivisions (departments) arranged down the rows.

Notice that *Headcount* and *Expenses* label two distinct sets of quantitative values, not subdivisions of a single category. Here's the same information, this time with the categorical subdivisions laid out across the columns:

Dept	Finance	Sales	Ops	Total
Headcount	26	93	107	234
Expenses	202,202	983,393	933,393	$2,118,988

FIGURE 4.7 This table is structured unidirectionally with the categorical subdivisions arranged across the columns.

These are *simple* unidirectional tables, in that they only contain a single set of categorical subdivisions, but tables can also have multiple sets, as in the following example:

Dept	Expense Type	Expenses
Finance	Compensation	160,383
	Supplies	5,038
	Travel	10,385
Sales	Compensation	683,879
	Supplies	193,378
	Travel	125,705
Total		$1,178,768

FIGURE 4.8 This table is structured unidirectionally, consisting of two sets of categorical subdivisions: departments in the left column and expense types in the middle column.

Because the categorical subdivisions are arranged in only one direction, in this case across the columns, this table is still structured unidirectionally.

Bidirectional

When a table is structured bidirectionally, more than one set of categorical subdivisions is displayed, and the sets are laid out both across the columns and down the rows. This arrangement is sometimes called a *crosstab* or a *pivot table*. The quantitative values appear in the body of the table, bordered by the categorical subdivisions, which run across the top and along the left side. This type of table can be best understood by looking at an example. Here's the same information found in the previous unidirectional table, but this time it is structured bidirectionally:

| Expense Types | Departments | | |
	Finance	Sales	Total
Compensation	160,383	683,879	844,262
Supplies	5,038	193,378	198,416
Travel	10,385	125,705	136,090
Total	$175,806	$1,002,962	$1,178,768

FIGURE 4.9 This is an example of a bidirectionally structured table.

In order to locate the quantitative value associated with a particular department and a particular expense type, you look where the relevant column and row intersect.

One of the advantages of bidirectional tables over unidirectional tables is that they can display the same information using less space. Here are the same examples of the previous unidirectional and bidirectional tables, but this time I've included a complete set of grid lines to make it easy to see the amount of space used by each. Here's the unidirectional version:

Dept	Expense Type	Expenses
Finance	Compensation	160,383
	Supplies	5,038
	Travel	10,385
Sales	Compensation	683,879
	Supplies	193,378
	Travel	125,705
Total		$1,178,768

FIGURE 4.10 This is an example of a unidirectionally structured table that contains two sets of categorical subdivisions: departments and expense types.

Here's the bidirectional version:

Expense Type by Dept	Finance	Sales	Total
Compensation	160,383	683,879	844,262
Supplies	5,038	193,378	198,416
Travel	10,385	125,705	136,090
Total	$175,806	$1,002,962	$1,178,768

FIGURE 4.11 This is an example of a bidirectional table that contains the same information as in *Figure 4.10* above but uses less space and displays additional values.

Count the cells (i.e., the individual rectangles formed by the grid lines) in each table. You'll find that the unidirectional table contains a grid of three columns by eight rows, totaling 24 cells, and the bidirectional table contains a grid of four columns by five rows, totaling 20 cells. That doesn't seem like much of a difference until you notice that the bidirectional version actually contains totals for each expense type in the right column that are not included in the unidirectional version, so not only is it smaller, it contains more information.

Table Design Solutions

As we explored the structural variations of tables, you may have noticed an affinity between the relationships that can be displayed in tables and the two primary ways to structure tables. This affinity indeed exists and is worth noting. Below is a table that lists the five types of relationships down the rows and the two structural variations across the columns, with blank cells intersecting these rows and columns. Take a few minutes to think about each type of relationship and fill in the blanks with either *Yes* or *No* to indicate whether the structural variation can effectively display the corresponding relationship. If the structural type will only work in certain circumstances, note them as well. You may find it helpful to construct sample tables and arrange them in various ways to explore the possibilities.

Relationship	Unidirectional	Bidirectional
Between a single set of quantitative values and a single set of categorical subdivisions		
Between a single set of quantitative values and the intersection of multiple categories		
Between a single set of quantitative values and the intersection of hierarchical categories		
Among a single set of quantitative values associated with multiple categorical subdivisions		
Among distinct sets of quantitative values associated with the same categorical subdivision		

. . . .

I'm confident that you've used your developing design skills effectively to identify solutions in the above exercise, but just so you have a nice, neat reference, I've included one in the *Summary at a Glance* on the next page.

Now you have a conceptual understanding of tables that includes what they are, when to use them rather than graphs, what they're used for, how they're structured, and which type of structure works for which type of quantitative relationship. Later, in the chapter on table design, we'll examine the specific design choices available to you to make quantitative messages in tables as clear and powerful as possible.

Summary at a Glance

Quantitative-to-Categorical Relationships

	Structural Type	
Relationship	*Unidirectional*	*Bidirectional*
Between a single set of quantitative values and a single set of categorical subdivisions	**Yes**	Not applicable because there is only one set of categorical subdivisions
Between a single set of quantitative values and the intersection of multiple categories	**Yes**. Sometimes this structure is preferable because of convention.	**Yes**. This structure saves space.
Between a single set of quantitative values and the intersection of hierarchical categories	**Yes**. This structure can clearly display the hierarchical relationship by placing the separate levels of the hierarchy side by side in adjacent columns.	Yes. However, this structure does not display the hierarchy as clearly if its separate levels are split between the columns and rows.

Quantitative-to-Quantitative Relationships

	Structural Type	
Relationship	*Unidirectional*	*Bidirectional*
Among a single set of quantitative values associated with multiple categorical subdivisions	**Yes**	**Yes**. This structure works especially well because the quantitative values are arranged closely together.
Among distinct sets of quantitative values associated with the same categorical subdivision	**Yes**	Yes. However, this structure tends to get messy as you add separate sets of quantitative values.

5 FUNDAMENTAL VARIATIONS OF GRAPHS

Different types of quantitative relationships require different forms of graphs. This chapter explores the fundamental variations of graphs that correspond to different quantitative relationships and then teams these variations with the visual components and techniques that can be used to communicate them most effectively.

Graphs always display information about relationships. The strength of graphs is their exceptional ability to present complex relationships so that we can see these relationships quickly and easily. The visual nature of graphs imbues them with this power. In this chapter we will first examine the methods that graphs use to visually encode data, both quantitative and categorical; then, we'll identify each type of relationship that graphs can display, and we'll finish by exploring the structural forms that can be used most effectively to display each relationship.

Encoding Data in Graphs

Think of yourself as a craftsperson, such as a carpenter, in the midst of your apprenticeship. It's my job to introduce you to the craft, helping you learn the required knowledge and skills until they become second nature and you can do the work confidently on your own. At the moment, I'm going to add a few simple tools to your tool belt and make sure that you know when and how to use them.

Graphs consist of several components (scales on axes, grid lines, bars, legends, etc.). Some components represent quantitative values (lines, bars, etc.), some represent categorical subdivisions, and some play a supporting role. The structural variations of graphs are defined primarily by differences in the components that encode quantitative values (e.g., lines versus bars).

Graphical Objects Used to Encode Quantitative Values

Quantitative values can be encoded in graphs using any of the following objects:

- Points
- Lines
- Bars
- Shapes with 2-D area

These objects vary in their ability to display different types of quantitative information.

POINTS

By a *point*, I mean any simple small geometric shape that is used to mark a specific location on the graph. A point often consists of a simple dot. In the following example of a scatter plot, a separate point marks each quantitative value.

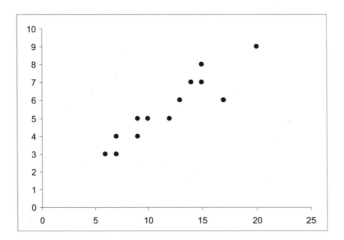

FIGURE 5.1 This graph uses points, in this case dots, to encode quantitative values.

The point closest to the top right represents a value of 9 on the Y axis and 20 on the X axis. In addition to a dot, any simple symbol (e.g., a small square or triangle) may be used to mark a point location on a graph.

Points are not necessarily restricted to scatter plots. Here's another example of a graph that uses points.

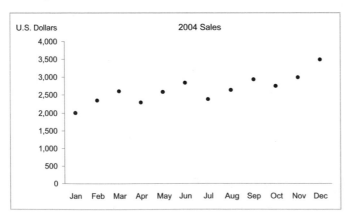

FIGURE 5.2 This graph uses points to encode quantitative values, this time in a form that is not a scatter plot.

This graph may seem a little unusual because you're probably accustomed to seeing graphs with information of this type use vertical bars to encode the quantitative values or data points connected by a line, as in the following example:

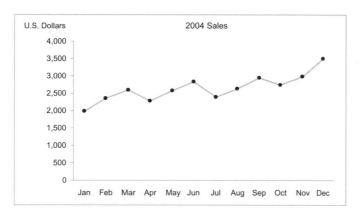

FIGURE 5.3 This graph employs two graphical objects, points and a line, to represent the same set of quantitative values.

In this case the graph uses two types of objects to encode the same set of sales values: points and a line. Sometimes the use of two graphical objects to encode the same set of quantitative values is not entirely redundant, as in this case where the points highlight the individual data values and the line helps us focus on the changes between values and the overall trend of those changes.

Points are seldom used alone without lines to connect them except in scatter plots where their small size and lack of direction (i.e., no definite up or down, right or left) elevates their ability to simultaneously encode quantitative values along both the X and Y axes.

LINES

A line is a collection of individual contiguous points extending in a single direction through space. Technically, a line is, by definition, always straight; a line's non-straight relative is called a *curve*. In common parlance, however, we think of lines as both straight and curved. There's no need for us to get overly technical, so we'll use the term *line* to refer to both shapes.

Lines are used in graphs to encode quantitative values in two ways:

- To connect individual data points (as in the example above)
- To display the trend of a series of data points (such as in the form of a trend line in a scatter plot)

When a line is used to connect individual values, it isn't necessary to display those values as points in addition to the line. Unlike in the example in *Figure 5.3*, which combines points and lines, the following example uses only a line:

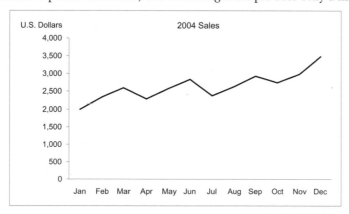

FIGURE 5.4 This graph uses only a line to encode the quantitative values.

A line that displays a trend is called a *trend line* or *line of best fit*. Here's an example of a trend line in a scatter plot:

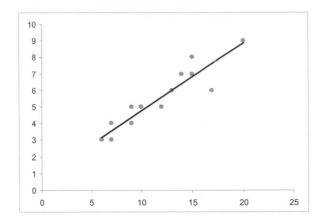

FIGURE 5.5 This graph uses points to represent individual quantitative values and a line to summarize the overall trend of those individual data points.

BARS

A bar can be thought of as thick line with the second dimension of width added to a line's single dimension of length. I realize that there actually is a name for the geometric shape of a bar—a rectangle—but I am intentionally downplaying the fact that bars have 2-D shape because the width of bars is not usually intended to convey any meaning in a graph. The quantitative information represented by a bar involves its length and its end point relative to the quantitative scale along an axis. So far, each of the graphical objects that we've examined has encoded quantitative values by means of their position relative to one or both axes. Tick marks (i.e., the little lines that mark the location of values along the axis, like the little lines on a ruler) and their numeric labels enable us to translate the position of data encoded as points, lines, or bars into quantitative values.

Take a moment to look at the bars on the graph below:

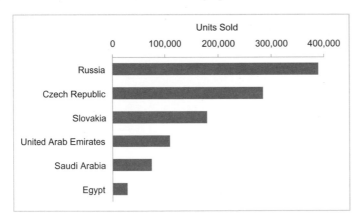

FIGURE 5.6 This graph uses bars to encode quantitative values.

The thickness of the bars has no meaning. I emphasize this because we have a tendency, thanks to the nature of visual perception, to assign greater value to things that are bigger, but nothing about a bar other than its length and the point at which it ends should influence our interpretation of its value. If the bar that represents the units sold in Slovakia were wider than the others, we would be tempted to assign it a value that is greater than its length alone suggests. Keep this in mind when you use bars in graphs, making sure that they are all of equal thickness so the reader's focus is drawn exclusively to their length for quantitative meaning. For most of us, this is not a concern, because software that generates graphs doesn't generally allow us to make some bars wider than others.

You may think it odd that the bars in the previous example run horizontally rather than vertically. Bars can run in either direction. There are times, however, when there is a clear advantage to using horizontal bars rather than vertical. We'll cover this topic later in the chapter on graph design.

Because the endpoints of bars encode quantitative values, points (e.g., dots) at the same locations as the bars' endpoints could replace the bars and convey the same meaning. So why use bars at all? Bars do one thing extremely well: due to their visual weight, they stand out so clearly and distinctly from one another that they do a great job of representing individual values discretely. In the previous example, the beginning of each bar at the axis that provides its categorical label (*Russia*, *Czech Republic*, etc.) clearly connects it with that categorical subdivision, and that subdivision alone. This makes it easy to focus on individual values and also to make clear comparisons between individual values, especially when they are right next to each other.

Because the length of a bar is one of the attributes that encodes its quantitative value, its base should always begin at the value zero. Notice how the lengths of the bars no longer support accurate comparisons of quantitative values in the graph below on the right, contrary to the accurate comparisons that are supported by the bars in the graph on the left.

Technically, bars that run vertically are called *columns*, and only the ones that run horizontally are called *bars*. However, because their purpose is exactly the same regardless of their orientation, I use the term "bar" to refer to both.

FIGURE 5.7 These graphs illustrate what happens when you use bars to encode data without beginning the quantitative scale at zero.

The only exception to this rule is when a bar is used to encode a range of values (sometimes called a range bar or floating bar), with the lowest value corresponding to its base and the highest value corresponding to its endpoint.

SHAPES WITH 2-D AREAS

The only other type of object that is commonly used to encode quantitative values is an assortment of 2-D shapes that represent value in proportion to their area (i.e., their two-dimensional size) rather than by their location on the graph. Here's an example in the form of a type of graph that has become quite familiar through frequent use:

This type of graph is called a *pie chart*. It is part of the larger family of *area graphs*,

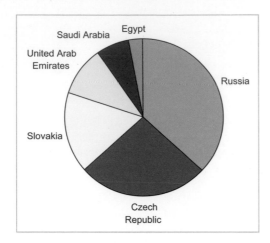

FIGURE 5.8 This graph uses 2-D area to encode quantitative values.

which use 2-D area to encode quantitative value. Pie charts use segments of a circle (a.k.a. slices of a pie) created by drawing lines from the circle's center to its perimeter. The size of each piece of the pie is equal to its value compared to the total value of all the slices.

You may be wondering, "where is the axis on this graph?" Good question. Remember, one of the defining characteristics of a graph is that it has at least one axis. Though it certainly isn't apparent, a pie chart does have an axis, but, unlike most graphs, this axis isn't a straight line. In pie charts, the perimeter of the circle serves as a circular axis. Just like any other axis, it could be labeled with measures of quantitative value corresponding to tick marks of equal distance all the way around the perimeter. This is seldom done, though, because it's hard to do without making the pie chart messy and difficult to read.

Speaking of "difficult to read," allow me to declare with no further delay that *I don't use pie charts*, and I strongly recommend that you abandon them as well. My reason is simple: pie charts communicate poorly. This is a fundamental problem with all types of area graphs but especially with pie charts. Our visual perception is not designed to accurately assign quantitative values to 2-D areas, and we have an even harder time when the third dimension of depth is added. Slices of a pie are particularly difficult to measure, but the 2-D areas of even less-challenging geometric shapes, like rectangles, are also hard to compare. If they're fairly close in size, it's difficult if not impossible to tell which is bigger, and when they're not close in size, the best you can do is determine that one is bigger than the other, but you can't judge by how much. In the next chapter, on visual perception, we'll examine the scientific evidence for this claim. For now, I simply want to point out that of the four graphical objects listed earlier as means to encode quantitative values, 2-D areas are not effective. Let's cross this value-encoding method from the list, which leaves us with only these three:

- Points
- Lines
- Bars

OTHER OBJECTS

On rare occasions other techniques are appropriate for encoding quantitative values (e.g., the varying sizes of circles in a *bubble chart*), but these occasions seldom arise in business communication, so we won't cover these other techniques in this book. Not only are these techniques rarely needed, they are also much harder to interpret than the ones we've looked at so far, requiring extraordinary care when you must use them.

People generally reach for these other techniques to display the correlation of more than two variables. In a normal scatter plot, you have two axes, and both provide the quantitative values for a single data point. Referring back to an earlier example, let's say you want to measure the correlation between employee tenure (i.e., years on the job) and employee productivity (e.g., the speed at which one can assemble widgets). In a normal scatter plot, you can display this relationship using the scale along one axis to represent tenure and the scale along the other to measure productivity. For each employee, you would place a point on the graph at the location where that employee's measures of tenure and productivity intersect. But what if you also need to account for the possible effect of compensation as a third variable? You could try to use a third axis, making the graph 3-D. This is a futile endeavor, even though it is supported by many software products, but the simulation of a third dimension on a flat piece of paper or computer screen results in a graph that is almost impossible to read. We'll look at this further in the later chapters on graph design.

Another method that is sometimes used involves different intensities of the same color (e.g., light gray to dark black) to represent a third dimension. So, in addition to the first two variables of tenure and productivity already mentioned, which are represented by location along the two axes, you could add the third variable by varying the color of the points (e.g., light dots to dark dots) to indicate varying amounts of employee compensation. As you can imagine, this is hard to interpret without a great deal of training and practice. Fortunately, there are usually better ways to handle more than two dimensions of information in graphs, which we'll cover in Chapter 11, *Design Solutions for Multiple Variables*.

Visual Attributes Used to Encode Categorical Data

The same objects that are used to encode quantitative values in graphs (points, lines, and bars) are also associated with categorical subdivisions. In a graph that displays sales in dollars (i.e., a measure of quantitative value) by quarter (i.e., a set of categorical subdivisions), something about each object that encodes a sales amount also needs to indicate the quarter it represents. In the following graph, the position of each bar along the X axis identifies the quarter:

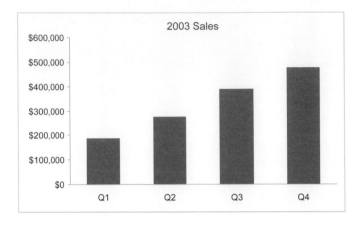

Categorical subdivisions may be represented in graphs by applying one of the following attributes to the objects that are also used to encode the quantitative values:

- Position (along an axis)
- Color
- Point shape
- Fill pattern
- Line style

POSITION

The most common attribute used to identify categorical subdivisions is position. In the following example, there is no question about which month is associated with each of the bars:

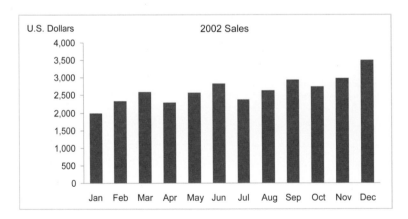

Position is the clearest means to associate quantitative values with categories in graphs. Position works equally well when the object that encodes the quantitative values is a point, line, or bar.

COLOR

The second most effective attribute for associating quantitative values with categorical subdivisions is color. In the next example, because the attribute of position has already been used to identify quarters, we must use a different attribute to add a distinction between the categorical subdivisions of *Direct* versus *Indirect* sales. Color will work just fine.

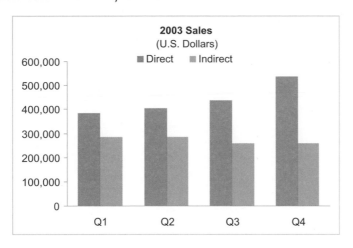

FIGURE 5.11 This graph uses two attributes for encoding categorical subdivisions: position along the X axis for the quarters of the year and color for the distinction between direct and indirect sales.

In this example, a legend has been added to label the two categorical subdivisions, *Direct* and *Indirect* sales. Interpreting categorical subdivisions that are encoded through the use of color is not quite as direct as interpreting data indicated by position, but once you've learned the legend and retained the association of blue with direct sales and beige with indirect sales in memory, you can easily interpret the graph. In fact, the colors are distinct enough that you can concentrate independently on the blue set of direct sales values or the beige set of indirect sales values with little distraction from the other set. Color works equally well as an attribute for points, lines, and bars, just as long as the object is not so small that you must strain to distinguish the colors. Some colors work better than others, which is a topic that we'll discuss in the next chapter on visual perception.

POINT SHAPE

This third attribute for encoding categorical subdivisions only applies when points are used to encode the quantitative values. Shapes such as dots, squares, triangles, asterisks, diamonds, dashes, etc. can be used to distinguish quantitative values that belong to different categories. In the graph below, both points and lines are used to encode the quantitative values, with differing symbols for the points to distinguish booking dollars from billing dollars:

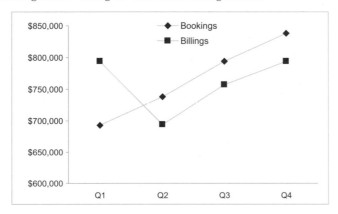

FIGURE 5.12 This graph uses differing symbol shapes to encode categorical subdivisions, in this case diamonds for bookings and squares for billings.

Points alone could have been used in this graph, but the addition of the lines makes it a little easier for our eyes to distinguish the two sets of points (one for bookings and one for billings) and to trace their trends through time. Differing point shapes, like the squares and diamonds in this graph, can be distinguished fairly easily but not as easily and rapidly as differing colors. Nevertheless, there are times when this method of encoding categorical subdivisions comes in handy, such as when you've already used position and color for other categorical distinctions, as in the following graph:

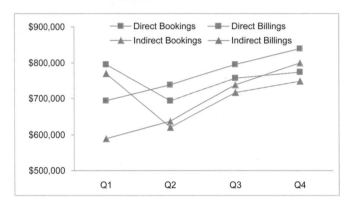

FIGURE 5.13 This graph uses position, color, and symbol shapes to encode categorical subdivisions, in this case position for quarters of the year, color for bookings vs. billings, and symbol shapes for direct vs. indirect.

FILL PATTERN

Fill patterns are only used to encode categorical subdivisions when the quantitative values are encoded as bars. Because bars have a 2-D area, this area can be filled with different patterns to distinguish categorical subdivisions. Here's a simple example:

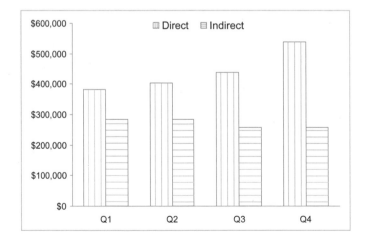

FIGURE 5.14 This graph uses fill patterns to encode categorical subdivisions, in this case horizontal and vertical fill lines.

Fill patterns are best used as a last resort. Not only are they harder for our eyes to distinguish than colors, they can play real havoc with our vision, causing a dizzying effect called the *moiré vibration*. This effect is especially strong if the patterns consist of vivid lines running in various directions, as in the next example:

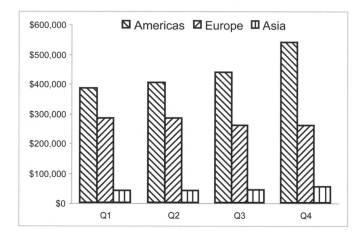

FIGURE 5.15 This graph uses fill patterns that are hard on the eyes and therefore difficult to read.

Try looking at this graph for a while, and you may actually feel ill, so it isn't hard to understand why fill patterns should be avoided whenever possible. If you must use them, do so with great care, muting the patterns and selecting ones that can be displayed together with minimal visual vibration. The primary reason that you may sometimes need to use fill patterns is if you must print your graph on paper or photocopy it for distribution, and color printing is not an option. In such cases, if you only need a few categorical distinctions, distinct shades of gray (e.g., black, dark gray, medium gray, and light gray), which are variations of color, are preferable to fill patterns.

LINE STYLE

Varying line styles can be used to encode categorical subdivisions when the quantitative values are encoded as lines. Lines may be solid, dashed, dotted, and so on. Here's an example:

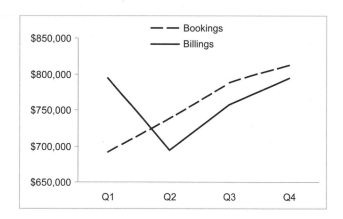

FIGURE 5.16 This graph uses line style to encode categorical subdivisions, in this case a solid and a dashed line.

Line styles don't work as well as colors because the breaks in the lines disrupt smooth perception of their flow. Consequently, like fill patterns, line styles are best reserved for occasions when color printing is not an option.

Relationships in Graphs

Graphs display relationships in quantitative information by giving shape to those relationships. Graphs vary primarily in terms of the types of relationships they are used to communicate, so it is useful to understand the specific types of relationships that graphs can display. With this knowledge, you will be able to quickly match the quantitative message you wish to communicate to the structural design that can do the job most effectively.

Essentially, there are seven types of relationships that business graphs are typically used to display:

- Nominal comparison
- Time series
- Ranking
- Part-to-whole
- Deviation
- Distribution
- Correlation

These seven types are easy to learn and differentiate.

Nominal Comparison

A *nominal comparison* relationship is the simplest of all. The categorical subdivisions have no particular order, for they represent a nominal scale. The goal is simply to display a series of discrete quantitative values so they can be easily seen and compared. "This is bigger than that," "this is the biggest of all," "this is almost twice as big as that," and "these two are far bigger than all the others" are some of the messages that stand out clearly in nominal comparison relationships. The relative amounts of employee turnover for a group of departments in a single month, quarter, or year is an example.

Time Series

A *time-series* relationship is nothing more than a series of relationships between quantitative values that are associated with categorical subdivisions of time, for example, years, quarters, months, weeks, days, hours, minutes, or even seconds. This type of relationship deserves to be treated separately because of the unique characteristics of time when displayed graphically. Time-series graphs are a powerful means of viewing trends as numbers march through time—a way to observe how something has changed over time. Time-series graphs do their job by displaying a single measure as it varies at each point in time along a sequential, chronological scale. No type of graph is used more commonly in business than the time series. A typical example is the closing price of stock for each day, spread over the last three months.

In order to determine whether your quantitative message is a time series, first state the message verbally, then look to see whether you used any of the following words:

- Change
- Rise
- Increase
- Fluctuate

The categorization of graphs into five out of seven of these relationship types was proposed by Gene Zelazny (2001) *Say It with Charts,* Fourth Edition. New York: McGraw-Hill. His taxonomy includes the following: 1) component comparison (i.e., part-to-whole), 2) item comparison (i.e., ranking), 3) time-series comparison, 4) frequency distribution comparison, and 5) correlation comparison. I've expanded this list to include nominal comparison and deviation relationships.

These words, with the exception of "trend," were suggested by Gene Zelazny (2001) *Say It with Charts,* Fourth Edition. New York: McGraw-Hill. The subsequent lists of words that accompany each relationship type below were derived primarily from the same source.

- Grow
- Decline
- Decrease
- Trend

If you did, then you should most likely display your message using a time-series graph.

Ranking

When graphs display *ranking* relationships, they communicate how the individual quantitative values associated with a set of categorical subdivisions relate to each other sequentially, sorted by size. The subdivisions are sequenced (e.g., 1, 2, 3 . . . or 1st, 2nd, 3rd . . .) either from lowest to highest or highest to lowest. A typical example is a ranking of customers based on the total dollar values of their purchases.

Words and phrases that suggest a ranking relationship include:

- Larger than
-
- Equal to
- Greater than
- Less than

Part-to-Whole

When graphs display *part-to-whole* relationships, they communicate how the individual quantitative values associated with a set of categorical subdivisions relate to the complete set of values and consequently to each other. This type of graph displays *ratios*. The common unit of measure is *percentage*. The whole is 100%, and each part is some relative portion of 100%. A typical example is the percentage of sales attributed to various sales regions relative to total sales.

Words and phrases that suggest a part-to-whole relationship include:

- Percent or percentage of total
- Share
- Accounts for X percent

Whenever your message uses one of these expressions, you are almost certainly communicating a part-to-whole relationship.

Deviation

When graphs display *deviation* relationships, they communicate how one or more sets of quantitative values differ from a primary set of values. They do this by expressing all quantitative values in relation to the primary set, specifically as a measure of the difference between them. Common examples include the following:

- The degree to which actual worker productivity differs from target productivity
- The degree to which sales over time differ from sales at some specific time in the past
- The degree to which headcount for each month differed from headcount for the previous month
- The degree to which sales of various products differ from sales of a particular product

Deviations are usually expressed using one of the following units of measure:

- Actual units (revenue dollars, number of employees, etc.)
- Ratios relative to the primary value (i.e., where the primary value always equals 1 or 100%, and all other values are expressed as relative rates or percentages)
- Positive or negative ratios relative to the primary value (i.e., where the primary value always equals 0 or 0%, and all other values are expressed as plus or minus rates or percentages)

Words and phrases that express deviation relationships include:

- Plus or minus
- Variance
- Difference
- Relative to

Distribution

Graphs that display distribution relationships describe how a set of quantitative values are distributed across its entire range, from the lowest to the highest. When a graph displays the distribution of a single set of values, the relationship is called a *frequency distribution,* for it shows the number of times something occurs (i.e., its frequency) within consecutive intervals of a larger numeric range. Graphs of this type visually display the same type of information that is communicated by the *range* and *standard deviation* measures of distribution. For instance, you may want to display the number of employees whose salaries fit into a consecutive series of salary ranges. It wouldn't make sense to count the number of employees for every distinct salary amount because there are far too many. Rather, you would break the entire range of salaries into smaller ranges, such as "less than $15,000," "greater than or equal to $15,000 and less than $30,000," "greater than or equal to $30,000 and less than $45,000," etc., and then count the number of employees that belong to each range.

Words that suggest a distribution relationship include:

- Frequency
- Distribution
- Range
- Concentration
- Normal curve, normal distribution, or bell curve

Correlation

Graphs that display a *correlation* communicate whether two paired sets of quantitative values vary in relation to each other, and, if so, in which direction (positive or negative) and to what degree (high or low). These graphs visually display the same type of information that is represented numerically by a *linear correlation coefficient*. A typical example is a correlation of marketing expenditures and subsequent sales. The existence of a correlation may indicate that one thing causes the other (i.e., a causal relationship) or that both things are caused by one or more other factors.

Words that indicate a correlation include:

- Relates to
- Increases with
- Decreases with
- Changes with
- Varies with
- Caused by
- Affected by
- Follows

It is a common mistake to assume that any correlation indicates causation. Just because two things vary together does not necessarily mean that one thing is acting on the other, causing the variation to occur.

Graph Design Solutions

You now know the visual objects and attributes that are available for encoding data in graphs, along with some of the strengths and weaknesses of each, as well as the types of relationships that graphs can be used to display. The next step is to learn the best structural solution for displaying each type of relationship. Let's examine the relationship types and think through the design solutions for each. Approaching the solutions in this manner will give you the tools that you need in a manner that deepens your understanding, enabling you to call them quickly to mind whenever needed.

Before we tackle the first type of graphical relationship, here's a reminder of the different objects that can be used to encode quantitative values:

Review each for a moment to remind yourself of their individual strengths for displaying quantitative messages.

FIGURE 5.17 This is a summary illustration of five objects (or combinations of objects) that can be used to encode quantitative values in graphs.

Nominal Comparison Designs

A nominal comparison graph displays a series of discrete quantitative values to highlight their relative sizes. Because the values are discrete, each relating to a separate categorical subdivision with no connection between them, you want to encode the quantitative values in a way that emphasizes their distinctness. Which visual encoding object emphasizes the distinctness of each value best? Bars do because they are quite distinct from one another, each carrying a great deal of visual weight. Graphs like the following example are quite common:

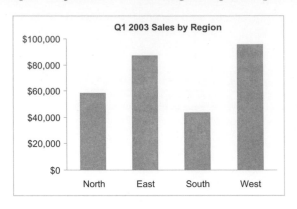

FIGURE 5.18 This is a nominal comparison relationship graph that displays sales by region.

Even though the bars in this example are vertical, horizontal bars would work just as well.

When bar graphs consist entirely of fairly long bars that are similar in length, it can be difficult to discern the subtle quantitative differences between them. Narrowing the quantitative scale so that it begins just below the lowest value and ends just over the highest value can help to make these subtle differences easier to see, but bars require a zero-based scale. In such cases, you can usually replace the bars with data points (e.g., dots), which are still able to display the values distinctly, but don't require a zero-based scale as bars do. Graphs of this type usually go by the name *dot plots*. Consequently, nominal comparison relationships can be effectively encoded using either of the following objects:

- Vertical bars
- Horizontal bars
- Data points

Time-Series Designs

A time-series graph displays quantitative values in relation to multiple, sequential points in time. Consequently, one axis of the graph needs to serve as the time scale, with labels for each unit of time (years, quarters, etc.). Values of time have a natural order. You would almost never display time other than in chronological order.

When you imagine a visual representation of time, how do you see it arranged on the page? Because of an age-old convention in most western cultures, there is only one way to lay out time on a page that wouldn't seem strange or confusing: horizontally, from left to right along the X axis. Given this fact, for time-series graphs we can eliminate one of the encoding methods shown in *Figure 5.17*. Take a look and determine which it is.

. . . .

Only horizontal bars cannot be used to display values of time. That's because horizontal bars use the X axis to label the quantitative scale, and the Y axis to label the subdivisions of time (e.g., months). Time-series graphs should always use the horizontal axis for the time scale and the vertical axis for the quantitative scale.

We've eliminated one graphical encoding method, which leaves us with four more from which to choose. Keep in mind as we continue that there may be more than one encoding method that works.

Vertical bars work in time-series graphs but should only be used when you wish to emphasize and compare the values associated with individual points in time rather than the overall pattern of values as they change through time. This is because the visual weight of the individual bars distracts somewhat from the overall shape of the data.

There is another method that we can eliminate that doesn't lend itself well to a time-series display. Looking at the available objects for encoding data once again, which do you think would not be an efficient means of displaying left-to-right quantitative values related to time?

If you imagine a graph that uses points alone to encode quantitative values across time, you will see that points, floating in space, don't help us visualize the sequential nature of time. Here's an example:

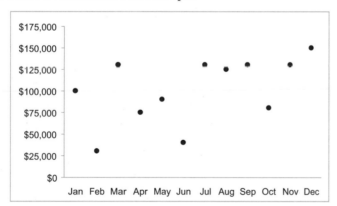

FIGURE 5.19 This graph uses points to encode quantitative values, which is not very effective for a time-series graph.

This graph doesn't give the sense of continuity that is required for displays of time, but adding lines to connect the points corrects this problem.

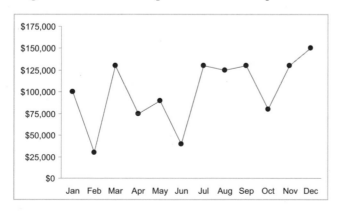

FIGURE 5.20 This time-series graph uses a combination of points and lines to effectively encode quantitative information.

With this simple change we have made the flow of time visible. Would we still have an effective display of time if we removed the points and used only the line? We would, but if we want to emphasize the individual values for each month along with their trend through time, using the points along with the line works best. By making one of these objects, either the points or the line, stand out more than the other, you can draw the reader's attention to either the individual values or to their overall flow through time.

We have identified three graphical objects that are appropriate for encoding quantitative values in time-series displays:

- Lines
- Points and lines
- Vertical bars

Ranking Designs

Ranking graphs display the order in which quantitative values associated with different categorical subdivisions relate to one another sequentially, from low to high or from high to low. One axis of the graph must provide a categorical scale and the other a quantitative scale. Because we want to emphasize each individual value and allow the reader to easily see its rank compared to that of any other value, we should encode the values using a graphical object that visually enforces the individuality of the values and their relative sizes. Which of the available graphical objects does this best? Bars do.

Here's a graph that uses bars, but so far it is only designed to display a nominal comparison relationship, because the categorical subdivisions are sorted alphabetically:

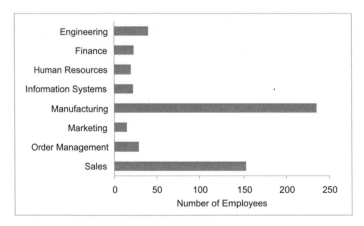

FIGURE 5.21 This graph uses bars to encode the values, which are the appropriate choice for a ranking display, but the bars are not arranged in a manner that highlights the ranking relationship.

Given our need to display a ranking relationship based on the number of employees associated with each department, this message would be displayed much more clearly if we arranged the departments in this order:

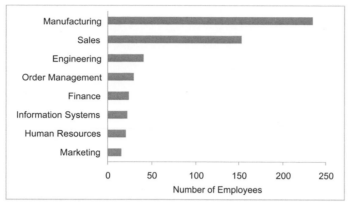

FIGURE 5.22 This ranking relationship is designed very effectively as a
bar graph with the bars sorted by size.

Now the ranking message is crystal clear.

Can you think of any reason why vertical bars, as opposed to the horizontal bars, would not work as well? Both types of bars work, but there are times when horizontal bars may be preferable to vertical bars, and vice versa. We'll take a look at such nuances later in Chapter 10, *Component-Level Graph Design*. For now, though, the following rules of thumb will come in handy:

Purpose	Sort Order	Bar Position
Highlight the highest values	Descending	Vertical bars: highest bar on left
		Horizontal bars: highest value on top
Highlight the lowest values	Ascending	Vertical bars: lowest bar on left
		Horizontal bars: lowest bar on top

In western cultures, we tend to think of the top, as opposed to the bottom, and the left, as opposed to the right, as the beginning. This convention is rooted in our written languages, which are read across the page from left to right and top to bottom.

Just as with nominal comparison relationships, bars can be replaced with data points to display a ranking relationship when it is useful to narrow the quantitative scale and in so doing remove zero from its base. There are therefore three graphical objects that we can use for effectively encoding quantitative values in ranking graphs:

- Vertical bars
- Horizontal bars
- Data points

Part-to-Whole Designs

As the name suggests, part-to-whole graphs relate parts of something to the whole. The best unit of measure is usually percentage, with the whole equaling 100% and each part equaling a lesser percentage corresponding to its value relative to the whole. In common practice, pie charts are often used to display part-to-whole relationships, but I've already explained my objections to the use of pie charts and other area graphs in the earlier section *Shapes with 2-D Areas*. A less objectionable structural solution than the pie chart is something called the *stacked bar graph*, but even this has its problems. Here are two examples of a stacked bar graph, one vertical and one horizontal:

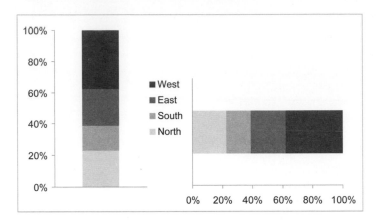

FIGURE 5.23 These two examples of part-to-whole relationships are both displayed as stacked bar graphs.

Rather than comparing them to determine which works best, take a moment to look for characteristics that hinder their effectiveness.

. . . .

Did you notice that it is easy to determine the percentage value associated with the north region but not as easy to determine the value of any other region? The value of the north region is easy because all you need to do is look at the percentage scale on the axis and the value is right there, but for the east region you must look at the values associated with the beginning and end of its portion of the bar (approximately 40% and 62%), then subtract the smaller from the larger value to get its percentage (approximately 22%). In other words, you have to do some math to determine the value. Now see if you can tell which has the larger percentage, the north or the east region. It's difficult to tell, because they appear to be about the same size. In fact, the east region is a little larger. It's hard to see this because of the way the north and east regions are stacked in the bar.

These problems can be solved without giving up the use of bars but instead by unstacking them. Look at these new graphs of exactly the same information, this time displayed using regular separate bars:

FIGURE 5.24 These two examples of part-to-whole relationships use separate bars to clearly display the contribution of each part to the whole.

This is better, isn't it? It is easy to interpret the percentage value of each region, and it is easier to tell that the east region has a slightly larger value than the north. It would be nice, however, if we could make it even easier to compare values that are close in size. What could we change about these graphs to accomplish this improvement? We could sort the regions in order of size, thereby placing those of similar size next to each other, which should make the comparison easier for the eye to detect. Let's try it:

FIGURE 5.25 These two examples display the same part-to-whole information that was displayed in *Figure 5.24*. This time the bars, each representing a different sales region, are sorted by size to make it easier to detect their relative sizes.

Now it's very easy to compare each region to the others. By sorting the regions by size, we've combined a display of ranking and part-to-whole into a single graph.

Bars, arranged either vertically or horizontally, allow you to clearly separate each categorical subdivision, such as the individual regions in the above graphs. Part-to-whole relationships, by their very nature, require this. Would lines or points offer a similar degree of distinction? No, lines wouldn't do this at all, and points, though individual, don't carry enough visual weight to display the relative sizes of the parts to the whole nearly as well as bars do.

Despite the disadvantages associated with stacked bars, one type of quantitative message justifies their use: when you wish to display the whole using an actual, non-ratio unit of measure (e.g., U.S. dollars) but also wish to provide some sense of the relative sizes of its parts. In the next example, the dominant message is total sales per quarter in U.S. dollars, but the use of stacked bars conveys the secondary message of relative regional sales as well.

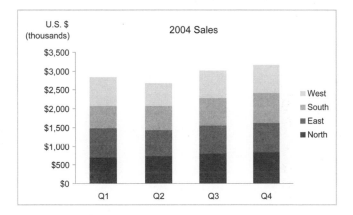

FIGURE 5.26 This is an example of a circumstance when the use of stacked bars is useful.

Even though the stacked bars can't communicate ideally, they work well enough here to provide additional useful information.

Our investigation leaves us with two variations of a single graphical object for encoding quantitative values in part-to-whole graphs:

- Vertical bars
- Horizontal bars

Deviation Designs

Deviation graphs display the degree to which one or more sets of quantitative values differ in relation to a primary set of values. Deviation relationships are often teamed with other relationships, especially time series. Displaying the difference between various measures and a reference measure over time is common practice in business. When combined with a time-series relationship, deviations are often encoded as lines to represent the continuous, flowing nature of time. When combined with other types of relationships (e.g., ranking and part-to-whole), or when functioning on their own, deviation relationships usually use bars.

Here's an example of a simple deviation relationship:

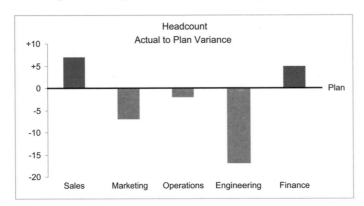

FIGURE 5.27 This deviation graph displays the difference between budgeted and actual headcount in actual numbers.

This first example displays the difference between planned headcount and actual headcount. Note that the positive variances, which in this case represent the instances you want to avoid (i.e., more employees than planned), are displayed in red to highlight them. The next example uses exactly the same data but this time expresses the variance as a percentage of the planned headcount:

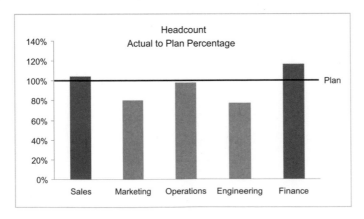

FIGURE 5.28 This deviation graph displays the difference between budgeted and actual headcount as a percentage.

Here's another example of the same information, this time displaying deviation as a percentage difference (positive or negative) from the plan.

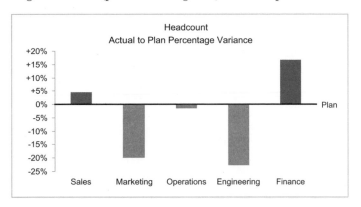

FIGURE 5.29 This deviation graph displays the actual headcount as a positive or negative percentage difference from planned headcount.

Let's look at one more example, this time combining a deviation relationship with a time-series relationship. Note that in this case a line is used to encode the values as they change through time:

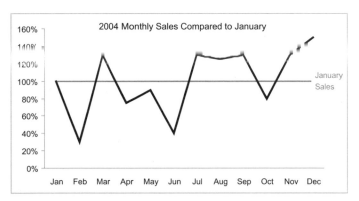

FIGURE 5.30 This combination of deviation and time-series relationships displays the variance of monthly sales to January sales, expressed as percentages.

If you wanted to display the relationship of each month's sales to those of the previous month, the graph would look something like this:

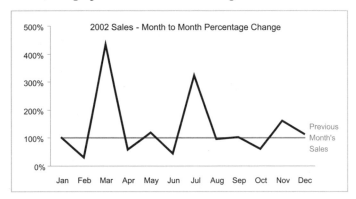

FIGURE 5.31 This combination of deviation and time-series relationships displays the variance between sales in a given month and the previous month, expressed as a percentage.

Our investigation has shown that deviation relationships can be effectively displayed using the following objects:

- Horizontal bars (except when combined with a time-series relationship)
- Vertical bars
- Lines
- Lines and points

Distribution Designs

Distribution graphs come in two basic types: 1) those that display the distribution of a single set of values, and 2) those that display the distributions of multiple sets of values so they can be compared. The best way to present a distribution depends on which of these two types you're attempting to display.

SINGLE DISTRIBUTION

Another name for a single distribution display is a frequency distribution. Frequency distribution graphs display how often something occurs, distributed across a series of consecutive, numeric ranges. The structural form that works most effectively to display a frequency distribution varies somewhat, depending on whether you wish to emphasize the number of occurrences in each range of the thing being measured or the overall shape of the distribution across the entire series of ranges. Given what you know about the unique strengths and weaknesses of each available object for encoding quantitative values, take a moment to determine which would work best for each of the following quantitative messages:

1. The number of orders that fall into each of the following size ranges:
    *   Less than $5,000
    *   Greater than or equal to $5,000 and less than $10,000
    *   Greater than or equal to $10,000 and less than $15,000
    *   Greater than or equal to $15,000 and less than $20,000
    *   Greater than or equal to $20,000 and less than $25,000
2. The time it takes to ship orders once they've been received, based on each of the following numbers of days:
    *   1 day
    *   2 days
    *   3 days
    *   4 days
    *   5 days
    *   6 days
    *   7 days
    *   8 days

· · · · · · ·

Given the emphasis in the first message on the frequency of occurrences in each of the order size categories, which graphical object would work best to highlight these individual measures? Both bars and points highlight individual values, but bars with their greater visual weight do a better job of making individual values stand out. The name for a graph that uses bars to display a frequency distribution is a *histogram*. Here's an illustration of this first scenario, displayed as a histogram:

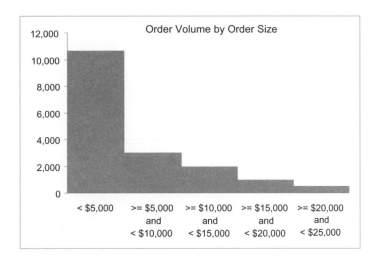

FIGURE 5.32 This graph uses vertical bars to display a frequency distribution, which is a form of graph called a *histogram*. The bars in a histogram do not conventionally have spaces between them, which visually suggests the continuous rather than discrete nature of the scale, with or without the intervening spaces.

Histograms are by far the most common structural solution for displaying frequency distributions. Although there is no inherent reason why horizontal bars could not be used just as effectively, histograms have become such a convention that it is best to stick with vertical bars.

Given the emphasis in the second message on the shape of the distribution rather than the values in the individual ranges, which graphical object would most effectively display this information? Lines actually draw the shape in simple visual terms. The name for a graph that uses a line to encode the shape of a frequency distribution is a *frequency polygon*. Here's an illustration of the second scenario, displayed as a frequency polygon:

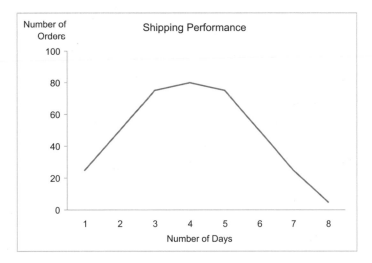

FIGURE 5.33 This graph uses a line to emphasize the shape of a frequency distribution, which is a type of graph called a *frequency polygon*.

In Chapter 2, *Numbers Worth Knowing*, I mentioned a name for a frequency distribution that has a shape similar to the one formed by the line in this graph; it is called a bell-shaped or normal curve.

Even though the shape of the distribution can be seen in a histogram as well, frequency polygons focus your attention exclusively on the shape by eliminating any visual component that would draw your eye to the values of the individual ranges.

By applying what we know about the graphical objects that can be used to encode quantitative values to the messages of frequency distributions, we've learned that two graphical objects can be used effectively in frequency distribution graphs:

- Vertical bars
- Lines

MULTIPLE DISTRIBUTIONS

It is often useful to display the distributions of multiple data sets in a single graph. You can combine multiple frequency polygons into a single graph, using separate lines to encode each data set. Here's an example of what this looks like:

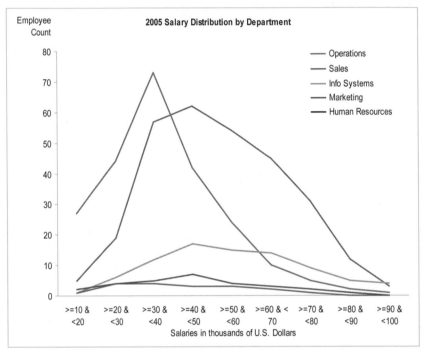

FIGURE 5.34: Lines can sometimes be used to display the distributions of multiple data sets in a single graph.

This works fine as long as there are no more than about five data sets. More lines than this would look cluttered and would be difficult to read.

When you need to display the distributions of more than a few data sets or to display the distributions of a series of data sets that each represent a different point in time along a time series, lines in the form of frequency polygons won't work. The solution to this situation was invented by John Tukey, a Princeton University statistician who contributed a great deal to the visual presentation of quantitative data, especially to support the exploration and analysis of data. Tukey's solution is called the *box plot* or *box-and-whisker plot*. What he calls a box is really just a bar—actually a range bar, which encodes a range (or distribution) of values from one end of the bar to the other. To this, however, he adds other objects to encode additional information about the shape of the distribution. These other objects include thin bars (which actually look more like lines) and data points in the form of a dot or short line running perpendicular to the bar. Here's an example of a simple box plot that displays the distribution of a single data set, with its components labeled:

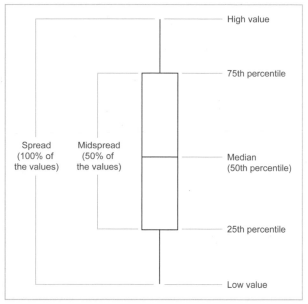

FIGURE 5.35 This is a box plot, which displays a single distribution of values, with each of its parts labeled.

As you can see, this simple combination of bars and data points communicates quite a lot about a distribution of values. Here's a list of the full set of facts that this elegant display features:

- The highest value
- The lowest value
- The range of the values from the highest to the lowest (called the *spread*)
- The median of the distribution
- The range of the middle 50% of the values (called the *midspread*)
- The point on or above which 25% and on or below which 75% of the values reside (called the *75th percentile*)
- The point on or above which 75% and on or below which 25% of the values reside (called the *25th percentile*)

Now let's look at the sample distribution displayed below to see what we can learn from it.

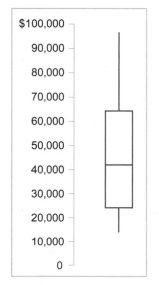

FIGURE 5.36 This example gives us a chance to practice making sense of box plots.

Assuming that this represents a distribution of salaries, the first thing it tells us is that the full range of salaries is quite large, extending from around $14,000 on the low end to around $97,000 on the high end. We can also see that more people earn salaries toward the lower rather than the higher end of the range. This is revealed by the fact that the median, encoded as the short horizontal line in the middle of the box at approximately $42,000, is closer to the bottom of the range than the top. Half of the employees earn between $25,000 and $65,000, which definitely indicates that this distribution is skewed toward the top end of the range. The 25% of employees who earn the lowest salaries are grouped closely together across a relatively small $10,000 range of salaries. Notice the large spread represented by the top 25% of the salaries. This tells us that as we proceed up the salary scale there appear to be fewer and fewer people within each interval along the scale, such as from over $60,000 to $70,000, from over $70,000 to $80,000, and from above $90,000 to $100,000. In other words, salaries are not evenly spread across the entire range; they are tightly grouped near the lower end and spread more sparsely toward the upper end where the salaries are more extreme compared to the norm.

When you need to display multiple distributions for comparison, box plots are hard to beat. Take a minute to study the example below to see what you can learn about how male and female salaries compare.

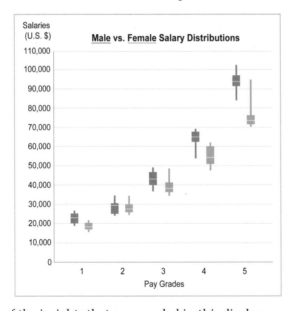

FIGURE 5.37 This box plot separately displays male and female salaries in five different pay grades.

Here are a few of the insights that are revealed in this display:

- On average, women are paid less than men in all salary grades.
- The disparity in salaries between men and women becomes increasingly greater as one's salary increases.
- Salaries vary the most for women in the higher salary grades.

These insights were submitted by Christopher Hanes in response to a data visualization competition that I judged for *DM Review* magazine, which used this graph.

To summarize, we've learned that multiple distributions can be displayed using the following objects:

- Lines
- Bars and points (i.e., boxes)

Correlation Designs

Correlation graphs display the relationship between two paired sets of quantitative values to communicate whether or not they are related, and, if so, the direction of the relationship (positive or negative) and the degree of the relationship (strong or weak). Because the relationship is between two sets of quantitative values rather than between categorical subdivisions and quantitative values, both the X and Y axes of the graph must provide quantitative scales. In this way we are able to display a single point on the graph for each pair of values as the intersection between a particular value on the X axis and a particular value on the Y axis. Let's say that you need to display the potential correlation between the heights of male employees and their salaries. In order to encode the values for an employee who is 70 inches tall and earns a salary of $60,000, you would find 70 on the Y axis, and then move to the right until you are in line with $60,000 on the X axis, and mark that spot on the graph, as illustrated below:

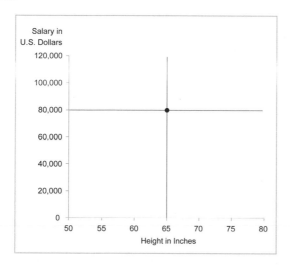

FIGURE 5.38 This illustrates the method used to position values of correlation on a graph with X and Y axes.

Based on this illustration, it doesn't take much imagination to recognize that points work perfectly for encoding correlation values. Here's a graph that displays an entire series of paired employee heights and salaries:

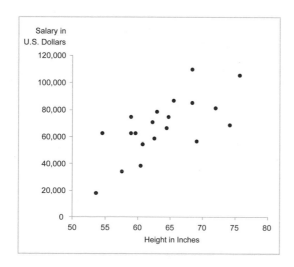

FIGURE 5.39 This graph displays the correlation between employees' heights and their salaries using points to represent the correlated values.

As pointed out in Chapter 2, *Numbers Worth Knowing*, there is a name for a graph
that uses points to mark values of correlation along the X and Y axes, both of
which display quantitative scales: *scatter plot*. Does the scatter plot above indicate
that there is a correlation between employee height and salary? What can we do
to make the potential correlation more visible? We can add a trend line.

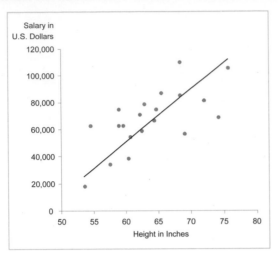

FIGURE 5.40 This scatter plot
includes a trend line to highlight
the overall pattern of correlation.

It is now easier to see by the slightly upward direction (from left to right) of the
trend line that there is a slight positive correlation between employee height and
salary, but it is fairly weak; the points are loosely grouped around the line.

Scatter plots are very effective for displaying correlations, with points to mark
the values and lines to highlight the pattern. We could end our examination of
structural solutions to the display of correlations right here, but a fundamental
problem crops up occasionally in business communications that motivates us to
consider additional solutions. Think about the audience for the graphs that you
prepare. Do they all know how to interpret scatter plots, and, if not, are there
those who are not willing to learn? Because you may not be able to rely on the
use of scatter plots in all circumstances, let's look at additional solutions.

Can you come up with a different way to display correlations? Use the
example that we've already been working with, the correlation of employee
heights and salaries. Can you somehow combine the components available for
use in graphs to display the correlation in a way that can be understood more
intuitively than a scatter plot?

· · · · · · ·

Were you able to come up with a viable solution? Even if you couldn't, I'm
sure the attempt was well worth the effort as a means to reinforce what you've
learned about graphs. I'm going to offer a solution that uses two sets of bars to
encode the two sets of paired quantitative values. Because we have two quantitative scales, inches for height and U.S. dollars for salary, we need two axes
to display them, but we can't use our normal arrangement of an X and Y axis
because it wouldn't work to display one set of bars going horizontally and one

going vertically. I want to position the bar that represents a particular employee's height near the bar that represents that same employee's salary so it's easy to see the relationship between the two values. There are two ways that we could structure this graph: one has the two scales and the two sets of bars running in the same direction, and one has them running in opposite directions.

Let's start with the version that has the two scales and sets of bars running in the same direction as this is closer to the norm. Because we can't display both scales on the same axis, we'll place two versions of the same axis, either vertical or horizontal, on opposite ends of the graph, either top and bottom or left and right. I'll use vertical axes to illustrate. Take a look at the following example, and see whether you can read it without instruction.

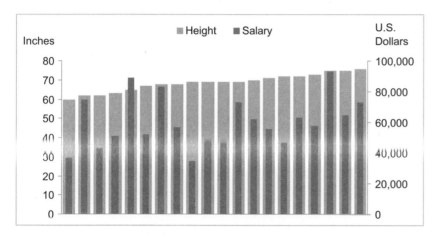

FIGURE 5.41 This graph uses bars to display correlations. This is not a conventional graph but one that I've designed for use when readers don't understand scatter plots. I call this a *Correlation Bar Graph*.

Because a Correlation Bar Graph is not a standard format included in software packages, Appendix D, *Constructing Paired Bar and Correlation Bar Graphs with Microsoft Excel*, gives instructions for constructing this type of graph using *Excel*.

Even though the two sets of bars overlap, each stands out as distinct because the bars representing the heights are wider and a different color than the bars representing the salaries. Our eyes can trace the pattern of either set without difficulty. Because our purpose is to show the correlation of salary to height, I've arranged the order of the bars in a particular way to make this correlation easier to detect. Can you see what I've done? The pairs of bars are presented in order of the employees' heights, from shortest to tallest. This makes it easier to see if there is a relationship between increasing height and salary. Does a correlation appear to exist? One thing that's clear is that the correlation certainly isn't perfect: as height increases, salary does not either increase or decrease in direct proportion. Don't make the mistake of reading too much into the actual heights of the *Salary* bars in relation to the *Height* bars. To focus on their correlation, you should simply see whether there is a tendency for the salary bars to fairly consistently either increase or decrease as the heights increase. If the two sets of bars tend to increase together, there is a positive correlation; if salaries tend to decrease as heights increase, there is a negative correlation. Just as with scatter plots, it would be easier to see the trend by using a trend line. Most software products that support graphing won't draw a single trend line that displays the relationship between two sets of bars, but if the software can draw trend lines, it is likely that it can draw one for each set of bars, which will serve our purpose. Take a look at the additional clarity that a set of trend lines provide:

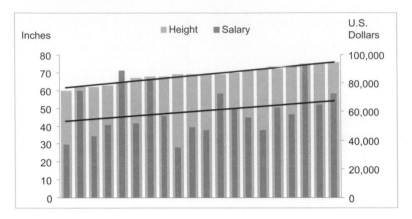

FIGURE 5.42 This *correlation bar graph* includes trend lines.

What the graph with trend lines reveals more clearly is that there is a tendency for salaries to increase as heights increase. The fact that the salary bars vary significantly relative to the salary trend line, however, tells us that the correlation is not a strong one.

We were able to reveal the same basic information about the correlation between employee heights and salaries in the above graph using vertical bars as we did in the earlier scatter plot. Scatter plots are superior overall, but if you suspect that your readers will struggle trying to understand a scatter plot, the correlation bar graph format may produce better results. If your data sets are much larger than those in the examples above, however, you may be forced to use a scatter plot. You can display much larger volumes of data in a smaller area with scatter plots than with correlation bar graphs.

An alternative to using scales and bars that run in the same direction is the *paired bar graph*. A scale on the left is read from right to left for one set of values, and a scale on the right is read from left to right for the other. Here's an example that displays the exact same employee height and salary information as before:

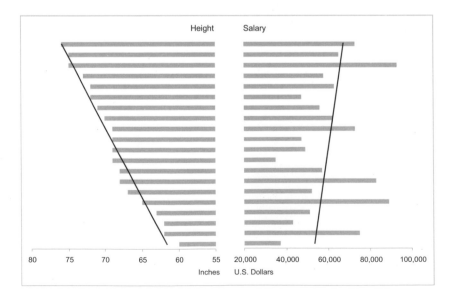

FIGURE 5.43 This is a *paired bar graph*.

Appendix D, *Constructing Paired Bar and Correlation Bar Graphs with Microsoft Excel*, gives instructions for constructing a paired bar graph using *Excel*.

Conventionally, the scales and bars in paired bar graphs run horizontally rather than vertically. It is certainly unusual for a scale to be arranged from right to left like the height scale in this graph, so if you use this type of graph you might want to give your readers a heads-up about how to read it, but the juxtaposition of two corresponding sets of paired values displays the message in a way that works.

Summary at a Glance

Relationship	Value-Encoding Object			
	Points	Lines	Points & Lines	Bars
Nominal Comparison	When there is a need to narrow the quantitative scale, and in so doing, remove zero from its base	Avoid	Avoid	Either horizontal or vertical bars
Time Series	Avoid	Categorical subdivisions on X axis, quantitative values on Y axis; emphasis on overall pattern	Categorical subdivisions on X axis, quantitative values on Y axis; equal emphasis on overall pattern and individual values	Categorical subdivisions on X axis, quantitative values on Y axis; emphasis on individual values
Ranking	When there is a need to narrow the quantitative scale, and in so doing, remove zero from its base	Avoid	Avoid	Either horizontal or vertical bars
Part-to-Whole	Avoid	Avoid	Avoid	Either horizontal or vertical bars
Deviation	Avoid	Especially useful when combined with time series	Useful when combined with time series and when a slight emphasis on individual values is desired	Either horizontal or vertical bars, except when combined with time series, which requires vertical bars
Distribution				
Single	Avoid	Known as a *frequency polygon*; emphasis on overall pattern	Avoid	Known as a *histogram*; emphasis primarily on individual values
Multiple	Use to mark the median in a box plot	Use for up to five distributions	Avoid	Use in the form of range bars in box plots
Correlation	Known as a *scatter plot*	Avoid	In this case the line is a trend line, not a line that connects the points.	Either horizontal or vertical bars; can be structured either as a *correlation bar graph* or a *paired bar graph*

PRACTICE IN SELECTING TABLES AND GRAPHS

Learning requires practice. Through practice you will reinforce what you've learned by embedding it more securely in your memory and strengthen your ability to make connections between the concepts we've examined and their application to the real world.

You may be tempted to skip this section of practice exercises, but I encourage you to take a few minutes to work through them. Taking these few extra minutes now to strengthen and deepen what you've learned may save you countless hours and a great deal of frustration over the course of your lifetime. This section consists of six business scenarios. Each requires that you make choices to determine the most effective design for communicating a quantitative message. Your choices for this set of exercises involve the following:

- Should the message be presented in the form of a table or a graph?
- If a table, which kind of relationship?
  - Between a single set of quantitative values and a single set of categorical subdivisions
  - Between a single set of quantitative values and the intersection of multiple categories
  - Between a single set of quantitative values and the intersection of hierarchical categories
  - Among a single set of quantitative values associated with multiple categorical subdivisions
  - Among multiple sets of quantitative values associated with the same categorical subdivision
- If a graph, which kind of relationship does it need to display?
  - Nominal comparison
  - Time-series
  - Ranking
  - Part-to-whole
  - Deviation
  - Frequency distribution
  - Correlation
- If a graph, which object or combination of objects for encoding the quantitative values would work best?
  - Points
  - Lines
  - Points and lines
  - Bars

Space has been provided in the right margin for your answers to these questions for each of the six business scenarios. Do your best to think through each scenario and respond without going back to review the contents of the chapter. Don't hesitate to review the chapter if you find yourself struggling, but allow yourself to struggle a bit first. You want to reach the point where the information resides in your head and is thoroughly understood.

Scenario #1

You are a financial analyst who works for the new Chief Financial Officer (CFO). You've spent the last month providing a spectrum of reports to help her become familiar with the company's financial state. She has come to believe that expenses are excessive, so she has scheduled a series of meetings, one with each department head, to discuss the problem and explore possible remedies.

She would like you to provide a single report that includes, by department, the headcount and expenses to date for the current quarter compared to budgeted headcount and expenses. This will give her the basic information that she'll need for each of the meetings. It's up to you to provide this in a manner that will serve her purpose most effectively.

Responses to Scenario #1:
Table or graph?

If a table, which kind?

If a graph, what kind of relationship?

If a graph, which graphical objects for quantitative encoding?

Anything else?

Scenario #2

You work as a product marketing manager for a company that manufactures and sells five distinct lines of software: 1) business productivity, 2) educational, 3) games, 4) programming, and 5) utilities. During the past five years, the relative amount that each has contributed to overall revenue has shifted. Five years ago your programming products were on top, but today they are dead last, and games are on top. As you examine the relative sales of each product line for each of the last five years, you notice a clear decline in the success of products that are more technical in nature (i.e., programming and utilities) and an increase in those that are geared toward entertainment.

You are preparing strategic recommendations for the company's five-year plan. To set the stage for your recommendation that the programming and utilities product lines be sold off and that the educational and games lines be expanded, you need to clearly present the shift that you've observed during the past five years. What form will you use to the present this observation?

Responses to Scenario #2:
Table or graph?

If a table, which kind?

If a graph, what kind of relationship?

If a graph, which graphical objects for quantitative encoding?

Anything else?

Scenario #3

Six months ago you developed and began teaching a new course entitled *Ethical Management*. When you initially proposed the idea for the class, your director was a little apprehensive about how well it would be received, but your past successes encouraged him to give you a shot. Now that you've been teaching it for a while and have worked the bugs out, it's time to give your director some evidence that he made the right decision in trusting your judgment and ability.

You've taught the course four times during the past month to a total of 100 students. Each student filled out an evaluation form at the end of the class, and you've tabulated the results. On a rating scale of 1 to 5, with 1 representing *worthless* and 5 representing *excellent*, the median rating for the course is 4, and the mean is 4.3. These ratings are exceptional. Not only is the average rating high, the range of ratings is tightly grouped around the ratings of 3, 4, and 5, with very few ratings of 2 and none of 1. When you compare these ratings to those that you received for another popular class that you also teach, their averages were about the same, but the spread of ratings for this other class were more broadly distributed, indicating that it doesn't work for all students as well as your new course.

You want to give this information to your director in a form that he will grasp with little difficulty. Once before, when you tried to communicate differences in the range of ratings between classes using standard deviations, you could tell that the director didn't really understand how to interpret them but was too embarrassed to admit it. This time you're going to approach the task differently. What form will your presentation take?

Responses to Scenario #3:
Table or graph?

If a table, which kind?

If a graph, what kind of relationship?

If a graph, which graphical objects for quantitative encoding?

Anything else?

Scenario #4

You have been promoted from Director of Customer Service to Vice President of Services. Before you were able to move full time into your new position, you had to recruit someone to replace you as director.

Your company spreads the work of customer service across four different customer service centers, one in each of four major geographical regions. Customers are able to rate their experiences with the service centers by responding to surveys distributed via email. Because you want the new director to focus on improving the centers that are scoring lowest in customers' ratings, you need to provide her with the mean rating of service for each service center during the most recent quarter. In what form will you present these summarized ratings?

Responses to Scenario #4:
Table or graph?

If a table, which kind?

If a graph, what kind of relationship?

If a graph, which graphical objects for quantitative encoding?

Anything else?

Scenario #5

You've been given a contract by a large manufacturing facility to analyze worker productivity data to see whether you can identify the cause of a recent decrease in productivity. What you learn from the new Operations Manager is that no matter how many additional people he hires, the result is reduced productivity. When the Operations Manager was hired six months ago, the General Manager told him that productivity had remained flat for years, and it was his job to increase it by 20% during the coming year. So far it has actually decreased by 20%.

After hearing this summary from the Operations Manager, one of the first things you decide to examine is the possible connection between staff additions and productivity decreases. Given your years of experience as a productivity analyst, you are not surprised to discover that increases in staff were proportionally related to decreases in productivity. You suspect that the addition of workers without changing anything else about the manufacturing process or facilities may have resulted in people simply getting in each other's way.

Responses to Scenario #5:
Table or graph?

If a table, which kind?

If a graph, what kind of relationship?

If a graph, which graphical objects for quantitative encoding?

Anything else?

You decide to show the Operations Manager the strength of this relationship of increased staff to decreased productivity before taking any further steps. You have monthly headcount and productivity statistics for the last year. Both headcount and productivity remained fairly steady until just after the Operations Manager's arrival. In what form will you present your information?

Scenario #6

For the first time ever, your company has built a database that contains comprehensive and reliable information about sales. Since it became available, you've been slicing and dicing the information in various ways, looking for answers to important questions that you've never before been able to investigate. One of your queries involved a list of every single order for the past year, sorted by size in U.S. dollars from the biggest to the smallest. You took your list and divided it into 10 equal groups of 10,000 orders each and labeled the groups *Top 10%*, *Greater than 10% through 20%*, and so on, to the final one labeled *Bottom 10%*. Next, you calculated the running percentage of total revenue associated with the orders, beginning with the largest order and continuing all the way to the smallest. You were then able to easily see the amount of revenue that each group of orders contributed to overall revenue.

You were amazed to discover that the top 10% of your orders contributed 87% of your total revenue. After the top 10%, the revenue contribution of the remaining 90% of your orders dropped off dramatically, with the last 50% contributing only 1% of total revenue. You have no doubt that executive management will find this discovery enlightening. You want to present this message as concisely and clearly as possible. You realize that if you don't hit them between the eyes with this important revelation in a single page of information, they won't bother reading it. What form will you give to this information to ensure that it hits the mark?

Responses to Scenario #6:
Table or graph?

If a table, which kind?

If a graph, what kind of relationship?

If a graph, which graphical objects for quantitative encoding?

Anything else?

You can find answers to the six scenarios in Appendix E, *Answers to Practice in Selecting Tables and Graphs.*

6 VISUAL PERCEPTION AND QUANTITATIVE COMMUNICATION

Quantitative communication, especially in the form of graphs, is predominantly visual. Thanks to science, how you see is fairly well understood, from the initial stimulus that enters your eyes to the interpretation of the information in the gray folds of your visual cortex. By understanding visual perception and its application to the communication of quantitative information in particular, you will learn what works, what doesn't, and why. This chapter brings the principles of visual design for communication alive in ways that are practical and can be applied skillfully to real-world challenges.

Mechanics of sight
Attributes of preattentive processing
Application of visual attributes to design
Gestalt principles of visual perception

Vision, of all our senses, is the most powerful and efficient channel for receiving information from the physical world. Approximately 70% of the sense receptors in our bodies are dedicated to vision. There is an intimate connection among seeing, thinking, and understanding. It is no accident that the terms we most often use to describe understanding are visual in nature, such as *insight*, *illumination*, and *enlightenment*. "I see" is the expression we use when something begins to make sense. We call people of extraordinary wisdom *seers*.

Graphs, and, to a lesser extent, tables, are visual means of communication. Although tables are visual in that we receive their information through our eyes, their reliance on language (i.e., text) reduces the degree to which they take advantage of the power of visual perception. Nevertheless, inattention to the visual design of tables can greatly diminish their effectiveness. Graphs are primarily visual in nature; the power of graphs and the power of visual perception are intimately linked.

Equipped with a fundamental understanding of visual perception—how it works, what works, what doesn't and why—your understanding of table and graph design will grow far beyond the memorization of a set of design principles. You will be able to apply these insights to new design challenges that extend well beyond the examples found in this book.

Colin Ware's book *Information Visualization: Perception for Design* is the best resource I've found for understanding visual perception as it relates to information design. Here's how he expresses its significance:

Why should we be interested in visualization? . . . The visual system has its own rules. We can easily see patterns presented in certain ways, but if they are presented in other ways, they become invisible . . . When data is presented in certain ways, the patterns can be readily perceived. If we can understand how perception works, our knowledge can be translated into rules for displaying information. Following perception-based rules, we can present our data in such a way that the important and informative patterns stand out. If we disobey the rules, our data will be incomprehensible or misleading.[1]

1. Colin Ware (2000) *Information Visualization: Perception for Design.* San Francisco: Morgan Kaufmann Publishers, page xviii.

Mechanics of Sight

The process of seeing and making sense of what we see begins with sensation and progresses to perception. Sensation is a physical process, involving the receipt of a stimulus. Perception is a cognitive process, involving the interpretation of the physical stimulus in an effort to make sense of it. For a visual stimulus to occur, there must be light. To see something, there must be an object out there that is either itself a source of light (e.g., the sun, a light bulb, or a firefly) or a surface that reflects light (e.g., a piece of paper on which a table or graph is drawn).

The light enters our eyes, which contain nerve cells that are sensors designed specifically to absorb light and translate it into neural signals (i.e., the language of the nervous system). These neural signals are then passed via the optic nerve to our brains where they are processed in an effort to make sense out of what we've seen.

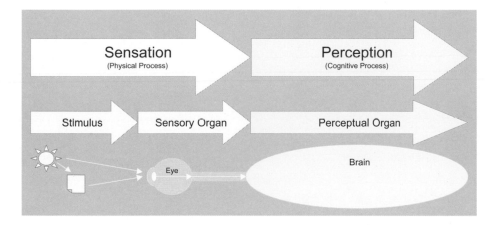

FIGURE 6.1 This diagram represents the process of visual perception.

The Eye

Our eyes operate like cameras, with several components working together to encode visual information. The surface of the eye, called the *cornea*, is a protective covering. Inside the eye, behind the cornea, resides the *iris*, a muscle that works like a camera shutter, enlarging or decreasing the size of the opening, called the *pupil*, to allow just the right amount of light to pass. The pupil is small when the amount of light is high (i.e., bright) and enlarged when there is less light, thereby allowing enough light through to provide a discernable image.

Behind the iris there is a *lens*, similar to the lens of a camera, which thickens to focus on objects that are near and thins to focus on distant objects.

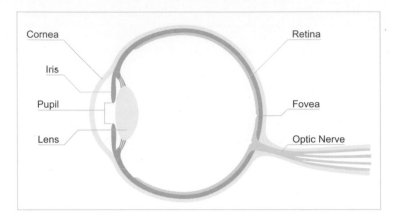

FIGURE 6.2 This is a diagram of the eye, with a label for each of the components that are significant to vision.

Eye diagram created by Keith Stevenson.

Once light passes through the pupil, it shines on the rear inner surface of the eye, called the *retina*. This is a thin surface coated with millions of specialized nerve cells called *rods* and *cones*. Rods and cones are two types of photosensitive (i.e., sensitive to light) receptors: rods specialize in sensing dim light and record what they detect purely in terms of black and white, and cones specialize in sensing brighter light and record what they detect in color. Cones are divided into three types, each of which specializes in detecting a different range of the color spectrum: blue, green, or red.

In the center of the retina is a small area called the *fovea* where the cones are densely concentrated. Images that are focused on the fovea are seen with extreme clarity because of the high number of cones there. When we want to examine something closely, we focus on it directly so that the light coming from it is directed to the fovea. When we focus in this manner on a printed page, we are able to see an amazing amount of detail, making distinctions between as many as 625 separate points of ink in a square inch.[2]

Rods and cones encode what they detect into neural signals, which are electro-chemical pulses, the language of the nervous system. These signals are then passed via the optic nerve to the brain where perceptual processing occurs. Millions of cones and rods work together simultaneously, each sending its signals to the brain at the same time through an extremely high-bandwidth channel at an amazing speed. Visual perception is the most highly developed and efficient of our senses.

The process of seeing is not smooth as you might imagine but consists of great rushes of information interrupted by frequent, brief pauses. This is because the fovea can only focus on a limited area at a time. Our eyes fixate on a particular spot and remain there for up to ½ second, sucking in the detail while at the same time we use what we are seeing in the non-foveal parts of the retina to survey the full field of vision for our next point of concentration. It takes our eyes about ¼ second to jump to the next point of fixation where the process is repeated. These little jumps are called *saccades*, and the motion is called *saccadic eye movement*.[3]

2. This statistic regarding the number of distinctions our eyes can detect in a square inch was derived from Edward R. Tufte (1983) *The Visual Display of Quantitative Information*. Cheshire CT: Graphics Press, page 161.

3. Much of the information about the saccadic nature of vision was derived from Marlana Coe (1996) *Human Factors for Technical Communicators*. New York: John Wiley & Sons, Inc.

As you are reading this page, your eyes are quickly jumping and fixating, jumping and fixating, taking in a few words at a time, moving from left to right across each line, then down to the next. During the brief moment when a saccade is occurring, when your eyes are jumping from one point of fixation to the next, you are not as sensitive to visual sensation as you are when your eyes are not moving. If something popped up on your computer screen during a saccadic eye movement and disappeared before the movement was complete, you would probably miss the fact that it was ever there.

Despite the millions of bits of information that our eyes are processing each second, we only see a little of what is out there, and only a fraction of that in detail. Some things grab our attention more than others.

The Brain

We don't see images with our eyes; we see them with our brains. Our eyes record light and translate it into electrical signals, passing them on to our brains, which is where images are perceived. What goes on in the brain produces what we experience as visual perception.

It is useful to think of the brain as a computer. Computers involve components and processes that are analogous to the components and processes of our brains. They receive input from a variety of devices (keyboard, mouse, digital camera, etc.). Let's trace the path of input from the keyboard as an example of how computers process input from the outside world, which is comparable to the sensory information we take in. When you tap the keys of your keyboard, electrical signals are initially routed into a special kind of memory called a *buffer*, in this case the keyboard buffer. This temporary location serves as a waiting room for the fraction of a second that the computer is busy until it is ready for fresh input. Once the computer is ready for the next input, these electrical signals move along the path into *working memory*, also called *random access memory (RAM)*. This is an extremely high-speed form of memory that the computer uses to process the information, doing what is necessary to interpret the message and respond to it. It is here that the bits and bytes are translated into meaningful terms, such as commands to do something in particular or words that are part of a document. If the information that the computer has perceived is meant to be kept around for future reference, it is then moved into some form of *permanent memory*, such as a hard disk. These three types of memory—buffer memory, working memory, and permanent memory—are analogous to three types of memory that process visual information in our brains. These are called:

* *Iconic memory* (a.k.a. the *visual sensory register*)
* *Short-term memory* (a.k.a. *working memory*)
* *Long-term memory*

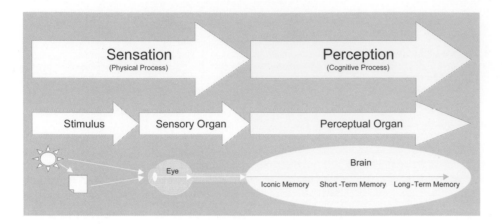

ICONIC MEMORY

Fresh electrical signals with encoded visual information from the eyes are routed through the optic nerve initially into iconic memory where each snapshot of input waits to be passed on to short-term memory. Information ordinarily remains here for less than a second while extremely rapid processing takes place before the information is forwarded to short-term memory. This part of perceptual processing is automatic and unconscious. For this reason it is called *preattentive processing*, as opposed to the conscious, higher cognitive processing that occurs later in short-term memory, which is called *attentive processing*. Preattentive processing is an extremely fast process of recognition and nothing more. Attentive processing, on the other hand, is a sequential process that takes more time and can result in learning, understanding, and remembering. Preattentive visual processing detects a limited set of attributes, such as color and the location of objects in 2-D space. Because preattentive processing is tuned to these attributes, they jump out at us, functioning as extremely powerful aspects of visual perception. If you want something to stand out in a table or graph as important you should encode it using a preattentive attribute that contrasts with the surrounding information. If you want a particular set of objects to be seen as belonging together as a group, you should encode them using a common preattentive attribute.

Preattentive visual attributes play an important role in visual design. They can be used to distinctly group and to highlight objects. We'll examine them in detail later in this chapter with special attention to how we can use our understanding of them and their role in visual perception to effectively design tables and graphs.

SHORT-TERM MEMORY

As visual information is moved from iconic memory into short-term memory, what our brains deem useful is combined into meaningful visual *chunks*. This form of visual information is used by the conscious, attentive process of perception that occurs in short-term memory. Two fundamental characteristics of short-term memory are:

- It is temporary.
- It has limited storage capacity.

Information remains in short-term memory from a few seconds to as long as a few hours if it is periodically rehearsed. Without rehearsal, it is gone after a few seconds. With the right kind of rehearsal, it is stored in long-term memory where it remains permanently.

Only somewhere around four chunks of information can be stored in short-term memory at any one time. For something new to be brought into short-term memory, something that is already there must either be moved into long-term memory or forgotten. Just like working memory in a computer, short-term memory stores information temporarily for high-speed processing. As this processing occurs, new sensations or memories are being moved into short-term memory while others are moved out either to long-term memory or to oblivion.

The most important thing to keep in mind about short-term memory is that readers of tables and graphs can only hold a few chunks of information in their heads at any one time. For instance, if you design a graph that includes a legend with a different color or symbol shape for 10 different sets of data, your readers will be forced to constantly refer back to the legend to remind themselves which is which because short-term memory is limited to about ˈɸ ɪ ɪ ɪ ɪ ɪ ɪ ɪ ɪ ɪ ɪ ɪ ˈlɪ ɪ ˈ ɪ ɪ ɪ ɪ ɪ ɪ ɪ ɪ ˈ ɪ ɪ ɪ ɪ ɪ ɪ ɪ ɪ ɪ ɪ ɪ ɪ ɪ in size. By designing a visual display of information to form larger, coherent patterns that combine multiple data into chunks, you can make it easier for your readers to stuff more information into short-term memory. This is one of the reasons that graphs are capable of communicating a great deal of information that can be perceived all at once while tables are limited to the purpose of look-up. We can't take a bunch of numbers from a table and chunk them together meaningfully for storage in short-term memory; we can, however, discern in a graph the image of a single, meaningful pattern that is made up of thousands of values.

LONG-TERM MEMORY

When we decide, either consciously or unconsciously, to store information for later use, we rehearse that chunk of information to move it from short-term to long-term memory. How we store the information involves an intricate network of links and cross-references, like indexes in a computer that help us to find information and retrieve it back into short-term memory when we need it. One chunk of information can have many links and cross-references.

Long-term memory is vitally important to visual perception because it holds our ability to recognize images and detect meaningful patterns, but we don't have to understand very much about long-term memory to become better designers of tables and graphs. What we mostly need to know is how to use preattentive visual attributes to grab and direct our readers' attention and how to work within the limits of short-term memory.

Evolution of Visual Perception

Like all other products of evolution, our visual perception developed as an aid to survival. Because of its evolutionary roots, visual perception is fundamentally

Colin Ware points out, in *Information Visualization: Perception for Design*, that our brains appear to have more than one type of *working* or short-term memory; each kind specializes in a different type of information processing. For instance, research has demonstrated that there are separate areas for visual and verbal information and that these areas don't compete with each other for space.

oriented toward action, always looking for what we can do with the objects we perceive. As designers of tables and graphs, we should understand how to use the attributes of visual perception to clearly communicate the affordances (i.e., the uses) of those tables and graphs, directing our readers to accurately and efficiently perform the act of interpretation, which they can then easily follow with appropriate actions.

Visual perception pays particular attention to the surfaces of objects. This ability has evolved to function preattentively, giving us clear and rapid awareness of the surfaces that surround us—an important survival mechanism. These are primal abilities that function below the level of consciousness. Let's identify and examine these attributes that put us in touch with the world visually, so that we can use them to communicate effectively.

This theory of visual perception's evolutionary functionality was first described by J. J. Gibson, who introduced the concept of *affordances* in 1979.

Attributes of Preattentive Processing

Preattentive processing is the early stage of visual perception that occurs below the level of consciousness at an extremely high speed and is tuned to detect a specific set of visual attributes. Attentive processing is conscious, sequential, and much slower. The difference between preattentive processing and attentive processing is easy to demonstrate. Take a look at the four rows of numbers below and determine, as quickly as you can, how many times the number 5 appears in the list:

> 987349790275647902894728624092406037070570279072
> 803208029007302501270237008374082078720272007083
> 247802602703793775709707377970667462097094702780
> 927979709723097230979592750927279798734972608027

FIGURE 6.4 This example demonstrates the slow speed at which we process visual stimuli that lack preattentive attributes.

How many? The answer is six. It took you some time to perform this task because it involved attentive processing. The list of numbers did not include any preattentive attributes that could be used to distinguish the 5 from the other numbers, so you were forced to perform a sequential search looking for the complex shape of the number 5. Now try it again, this time using the list below:

> 9873497902**7**5647902894728624092406037070**5**70279072
> 80320802900730**2**501270237008374082078720272007083
> 24780260270379377**5**709707377970667462097094702780
> 92797970972309723097**5**927**5**0927279798734972608027

FIGURE 6.5 This example demonstrates the fast speed with which we process visual stimuli that exhibit preattentive attributes.

This time it was easy because the fives were distinguished through the preattentive attribute of color. Because only the fives are black, and all the other numbers are light gray, the numbers that are black stand out in contrast. Given this example, it certainly doesn't take a great imagination to recognize the power of preattentive attributes used knowledgeably for visual communication.

Colin Ware has organized preattentive attributes into four categories: form, color, spatial position, and motion. I've reduced his list somewhat to serve our focus on the design of tables and graphs:

Colin Ware (2000) *Information Visualization: Perception for Design.* San Francisco: Morgan Kaufmann Publishers, pages 165 and 166.

Category	Attribute
Form	Orientation
	Line length
	Line width
	Size
	Shape
	Curvature
	Added marks
	Enclosure
Color	Hue
	Intensity
Spatial position	2-D position
Motion	Flicker
	Direction

In each illustration of a preattentive attribute below, only one object stands out as different from the rest, based on a variation of the attribute.

Attributes of Form

Attribute	Illustration	Attribute	Illustration
Orientation		Shape	
Line Length		Curvature	
Line Width		Added Marks	
Size		Enclosure	

Attributes of Color

Attribute	Illustration
Hue	
Intensity	

Hue is a more precise term for what we normally think of as color (red, green, blue, pink, etc.). Color is made up of three separate attributes; hue is one. Intensity applies to both of the other two attributes of color: *saturation* and *lightness*. In the illustration of intensity above, the object that differs is actually the same hue as the others (i.e., a shade of black), but it varies in intensity in that it is lighter. This three-value system of describing color is known as the *HSL* (hue, saturation, and lightness) system.

This system describes hue in units of degrees, from 0 to 360, along the circumference of a circle, illustrated by the following color wheel:

This system for describing color is sometimes called the *HSB system*, which stands for hue, saturation, and *brightness* rather than lightness.

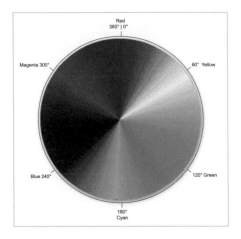

FIGURE 6.6 This is a color wheel, which represents hue in a circular manner.

Numbering starts at the top with red at 0° and continues around the wheel until it reaches the point where it started with red again, which is also assigned the value of 360°. If you could walk around this color wheel clockwise taking 60° steps, starting at red (0°), with each succeeding step you would be positioned at the following hues: yellow (60°), green (120°), cyan (180°), blue (240°), magenta (300°), then back to red again (360°).

Saturation measures the degree to which a particular hue fully exhibits its essence, ranging, for example, from fully red to paler and paler versions of red until the color no longer contains any of its essence and appears as white, as in the following example:

FIGURE 6.7 This is an illustration of the full range of saturation for the hue red, with 0% saturation on the left and 100% saturation on the right.

Saturation is expressed as a percentage, starting with 0% on the left, representing no saturation to 100% on the right, representing full saturation.

Lightness (or brightness) measures the degree to which a color appears dark or light, ranging from fully black to fully white. Here's an example of the hue red, with lightness ranging from fully dark on the left and fully light on the right:

FIGURE 6.8 This is an illustration of the full range of lightness for the hue red, with 0% lightness on the left and 100% lightness on the right.

Lightness is also expressed as a percentage, with 0% representing fully dark and 100% representing fully light.

Any color can be described through the use of these three measures: hue (0–360°), saturation (0–100%) and lightness (0–100%).

Attributes of Spatial Position

Attribute	Illustration
2-D Position	● ● ● ●

3-D position is also a preattentive attribute but it cannot be effectively applied to the design of tables and graphs rendered on a 2-D surface (e.g., a page or computer screen).

Attributes of Motion

Attribute	Description
Flicker	A characteristic of the object, such as a color, continues to go on and off, or the entire object may appear and disappear
Direction	The course of change in position of a moving object

Preattentive attributes of motion are very powerful attention-getters. Evolution has given us a heightened sensitivity to anything that suddenly appears in our field of vision. Ancestral memories of a saber-toothed tiger springing into the periphery of our vision still keep us vigilant. Despite the potential applications of preattentive attributes of motion to dynamic computer-based displays of tables and graphs, we'll be sticking to static displays in this book.

Application of Visual Attributes to Design

Preattentive attributes of visual perception can be used to make particular aspects of what we see stand out from the rest. Understanding these attributes enables us to design tables and graphs that clearly emphasize the most important information they contain, keeping the other visual components from drawing attention away from the key information.

Uses for Encoding Quantitative Values

Each of the attributes of preattentive processing can be expressed as values along a continuum. For instance, the preattentive attribute of line length can be expressed as a line of any length that the eye can see, not just simply as *short* and *long*. We perceive quantitative meanings in the variations of some visual attributes but not in others. Using the attribute of line length once again as an example, we naturally perceive long lines as representing a greater value than short lines. That is, we perceive the differing values of line length quantitatively. But what about an attribute like hue? Which represents the greater value: blue, green, red, yellow, black, or purple? Although it is true that each of these hues can be measured by instruments to determine its wavelength, and the number assigned to one hue would be greater or less than the number assigned to another, we don't think in these terms when we perceive hue. That is, we perceive different hues as different categorically; they're simply different, each belonging to a different expression of what we perceive as color. The other attributes of color, saturation and lightness, however, can be perceived quantitatively, based on more or less saturation and more or less lightness, which is why they are expressed as percentages rather than as degrees of angle around a circle, like hue.

This distinction between our perception of visual attributes as quantitative or categorical is significant. It tells us which attributes we can use to encode quantitative values in graphs and which can only be used to encode categorical subdivisions. Here's a list of the preattentive attributes that are useful in graphs with an indication of which can be perceived quantitatively:

Type	Attribute	Quantitatively Perceived?
Form	Line length	Yes
	Line width	Yes, but limited
	Orientation	No
	Size	Yes, but limited
	Shape	No
	Curvature	No
	Added marks	No
	Enclosure	No
Color	Hue	No
	Intensity	Yes, but limited
Position	2-D position	Yes

Although one might argue that we can also perceive orientation and curvature quantitatively, there is no natural association of greater or lesser value with these characteristics (e.g., which is greater, a vertical or a horizontal line?).

All of these preattentive attributes can be used to encode categorical subdivisions, including those that we perceive quantitatively. The most effective of these is 2-D position. For instance, the 2-D position of data points on a dot plot can simultaneously represent quantitative values based on the position of the data point relative to the Y axis and categorical subdivisions based on their position along the X axis, as in the following example:

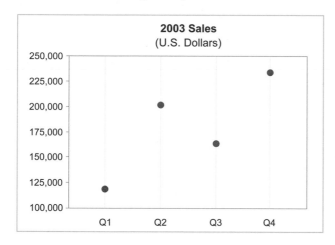

FIGURE 6.9 This dot plot uses the 2-D position of the data points to encode quantitative values along the Y axis and categorical subdivisions along the X axis.

The attributes that can be perceived quantitatively, but only in a limited way, are best reserved for encoding categorical subdivisions. This is because, even though we can tell that one value is greater than another, it is difficult to determine by how much. This is true of the attributes *line width* and *size* because both rely on our ability to assign values to the 2-D areas of an object, such as the slice of a pie chart or the width of a line. Although we can do a fairly good job of discerning that one object has a larger area than another, it is difficult to perceive by how much or to assign a value to the area. Take a look at the two circles below. I've assigned a value to the one on the left, and it is up to you to assign a value to the one on the right.

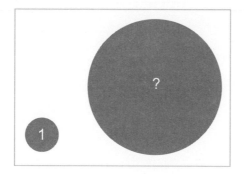

FIGURE 6.10 This figure illustrates the difficulty of perceiving differences in the 2-D areas of objects.

How much bigger is the large circle than the small circle? Given the fact that the size of the small circle equals a value of 1, what is the size of the large circle?

· · · · · · ·

The answer is 16. If you got the answer right, you have an unusual perceptual talent. Most of us have a tendency to underestimate the size of the larger object when faced with a task of this sort.

The color attribute of *intensity* poses a similar problem. For instance, we can use the color attribute of *lightness* as the measure of intensity, ranging from white (i.e., fully light) as the minimum value and black (i.e., fully dark) as the maximum along a gray scale. As long as there is enough difference in the amount of lightness in the colors of two objects, we can tell that one is lighter or darker than the other, but it is hard to assign a value to that difference. For this reason, avoid the use of color intensity as a means to encode quantitative value except in a pinch.

Perceptual Effects of Context

One fact that may surprise you about our perception of visual attributes such as color or size is that our visual senses are not designed to perceive absolute values but rather differences in values. Your sense of the brightness of a candle will differ depending on the context—i.e., in relation to the other values of brightness in your range of vision. If you light a candle when surrounded by bright sunshine, the candle won't seem very bright. If you light the same candle while standing in a pitch-black cellar, the candle will seem extremely bright in contrast to the darkness. That's because our visual receptors measure differences rather than absolute values. Let's consider another example, this time using hue as the visual attribute. Look at the two small rectangles that are in the two larger rectangles in this next example and compare their hues.

FIGURE 6.11 This illustration demonstrates the effect of context on the perception of hue.

How much darker is the small rectangle on the right than the small rectangle on the left?

. . . .

The truth is, there is no difference at all in their actual values when measured by an instrument. They appear different to us, however, because our perception of them is influenced by differences in what surrounds them.

Each attribute of visual perception is influenced by context, including attributes as seemingly straightforward as line length. Take a moment to examine the two pairs of lines below.

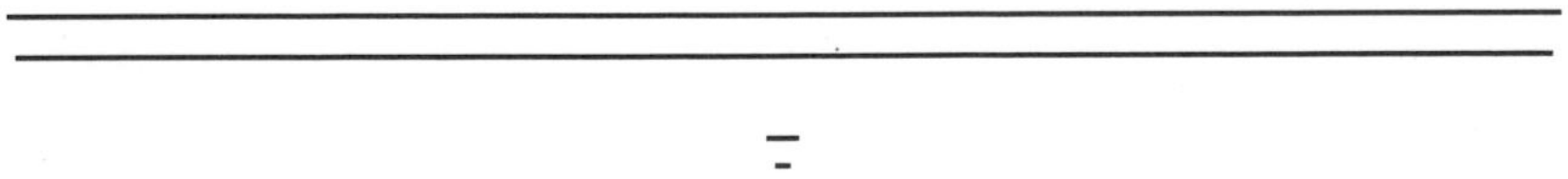

FIGURE 6.12 This figure demonstrates that we perceive relative differences in line length rather than absolute values of line length.

It appears that the difference in length between the top pair of lines is less than the difference in length between the bottom pair. In fact, the absolute difference in length between the two lines in each pair is exactly the same. We perceive differences in line lengths in terms of how the length of one line compares to the length of another, that is, as a ratio or percentage of one to the other, rather than in terms of absolute values. Because the lengths of the second pair of lines represent a ratio of 2 to 1, the difference appears greater than that between the first pair of lines, which has a ratio of 9.5 to 9.4.

Here's one more example of the effect of context on the perception of visual attributes. In this case the shapes at the ends of the two horizontal lines affect our perception of the lengths of those lines.

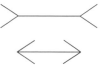

FIGURE 6.13 This figure demonstrates the effect of context on the perception of line length.

The horizontal line on the top is exactly the same length as the line on the bottom, but the diagonal lines at their ends influence our perception of their lengths.

Optical illusions such as these clearly demonstrate the effect of context on visual perception. Even when we know we are looking at an optical illusion, misperception persists, because visual perception measures differences in the attributes of what we see, not absolute values. It's up to us to use this knowledge when we are designing tables and graphs to prevent context from contributing to misperception of the message we are trying to communicate.

Some combinations of differing attributes of a visual object and its context work well to highlight the object, and some render it imperceptible. For instance, text is easiest to read when it is black on a white background. Other hue combinations are also acceptable, such as white text on a black background, and, to a lesser degree, red or dark blue text on a white background. In contrast, some combinations simply don't work, such as yellow on a white background or blue on a black background.

FIGURE 6.14 Some contrasting hues work well for clear perception, and some don't. You must look closely to see that there is any text at all in the lower right rectangle because blue does not stand out distinctly against a black background.

We could examine many more examples, both good and bad, but it's not necessary to memorize the combinations that do and don't work. As you can see from these examples, it's fairly easy to tell the difference when you take the time to notice. What's important is that you pay attention and use your own sense of visual perception to select combinations that are effective for clear and efficient communication and avoid those that aren't.

Limits to Distinct Perceptions

There is a limit to the number of visual distinctions of a single attribute that can be clearly discerned in a graph. For instance, if you create a line graph displaying departmental expenses across time, with a separate line of a distinct hue to represent each department in your company, readers of the graph will only be able to easily distinguish a limited number of hues. This is true for each visual attribute that can be used to encode categorical subdivisions, except 2-D position, because there is no limit to the number of different positions along the axis of a graph that are clearly distinguishable.

Colin Ware nicely discusses the limits on our ability to distinguish visual attributes:

> Pre-attentive symbols become less distinct as the variety of distracters increases. It is easy to spot a single hawk in a sky full of pigeons, but if the sky contains a greater variety of birds, the hawk will be more difficult to see. A number of studies have shown that the immediacy of any pre-attentive cue declines as the variety of alternative patterns increases, even if all the distracting patterns are individually distinct from the target.[4]

4. Colin Ware (2000) *Information Visualization: Perception for Design.* San Francisco: Morgan Kaufmann Publishers, page 167.

Ware reports research findings indicating that, when reading graphs, we can distinguish preattentively between no more than about eight different hues, about four different orientations, and about four different sizes, and that all the other visual attributes of preattentive perception should be limited to less than 10 distinct values as well. My own experience indicates that, with the exception of hue and shape (i.e., simple shapes that can be used as points on a graph, like squares, triangles, asterisks, dots, etc.), it is best to limit the number of distinctions for any one attribute to no more than four.[5]

5. Ibid. pages 195 and 196.

You may think that it would be worthwhile to use a larger number of distinct values even though this would force your readers make use of slower, attentive processing to interpret your graph. Unfortunately, if you do this, your readers will run up against the limits of short-term memory, which allows them to store the meaning of no more than four distinct values at a time.

Another limitation to keep in mind is that preattentive processing generally cannot handle more than one visual attribute of an object at a time. Let me illustrate. In the next figure, only the attribute of lightness is used to distinguish between three values, represented with black, dark gray, and light gray:

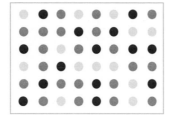

FIGURE 6.15 This figure illustrates the use of distinct values of a single visual attribute, in this case lightness, to encode objects as different from one another.

It is not difficult to pick out the black circles as distinct from those that are dark gray or light gray, the dark gray circles as distinct from those that are black or light gray, etc. Let's complicate it now by introducing a second preattentive attribute: by using two shapes (circles and squares), along with the three shades of lightness (black, dark gray, and light gray), we double the number of distinct values, but we do so at a cost. Take a look.

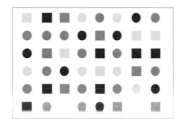

FIGURE 6.16 This figure illustrates the problem that results from encoding distinct values using more than one visual attribute, in this case lightness and shape.

It's still fairly easy to focus on just the squares, or just the circles, or just the black objects, or just the dark gray objects, or just the light gray objects, but try to pick out the dark gray squares or the light gray circles. The process of simultaneously focusing on multiple visual attributes (e.g., shape and lightness) requires slower, attentive processing. You can imagine how much more complicated it would get if we threw in a few more shapes or added the third attribute of size.

A related consideration is that when you select the values of a single attribute that you will use in a graph (e.g., multiple values of hue to encode categorical subdivisions), you must make sure that they are far enough apart from each other along the continuum of potential values to clearly stand out as distinct. Although I can choose up to four different shades of lightness, shades that are too close to one another won't work. Here are two sets of four distinct values, encoded as different shades of lightness. The four values of lightness on the left were carefully chosen from the full range of potential values for easy recognition of distinctness, but those on the right were not.

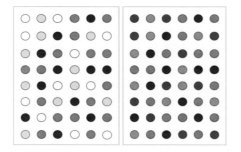

FIGURE 6.17 This figure illustrates the importance of selecting distinct values of a visual attribute that can be easily distinguished. The four shades of lightness encoded in the circles on the left are distinct, but the four on the right are not.

With a little effort, you can develop once and for all a set of distinct values for each of the visual attributes, which you can use over and over, rather than repeatedly going through the process of determining the distinctions that work.

Here's a list of 12 hues that are distinct enough to work well together:[6]

6. This list, with the exception of cyan resulted initially from research conducted by Brent Berlin and Paul Kay (1969) *Basic Color Terms: Their Universality and Evolution.* Berkeley: University of California Press. To their original list of 11 hues, Colin Ware (2000) added cyan in his book *Information Visualization: Perception for Design.* San Francisco: Morgan Kaufmann Publishers, pages 135 and 136.

1. Red
2. Green
3. Yellow
4. Blue
5. Black
6. White
7. Pink
8. Cyan
9. Gray
10. Orange
11. Brown
12. Purple

Each of these hues meets the requirement of distinctness, but the context of their use should sometimes lead you to reject some and select others. A good example, especially if you communicate with an international audience, involves cultural context. Particular hues carry meanings that vary significantly among cultures. In most western cultures, we think of red as signifying danger, warning, heat, and so on, but in China, red represents good fortune. The key is to be in touch with your audience and take the time to consider any possible associations with particular hues that might influence their interpretation of your message.

Various colors also affect us in different ways on a fundamental psychophysical level. Some colors are strong and exciting, grabbing our attention, and others are more neutral and soothing, fading more into the background of our attention. Edward Tufte suggests that colors found predominantly in nature, especially those that are on the light or pale side (e.g., shades of gray and light yellows, greens, and blues), are soothing and easy on our eyes.[7] Such colors are particularly useful in tables and graphs for anything that you don't want to stand out above the other content. On the other hand, fully saturated, bright versions of just about any primary hue tend to demand attention. These should rarely be used except when you intend to highlight information. Notice the difference between the pale shades on the left and the brighter shades on the right below:

7. Edward R. Tufte (1990) *Envisioning Information.* Cheshire CT: Graphics Press, page 90.

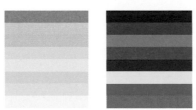

FIGURE 6.18 This figure shows subtle, soothing hues on the left, and fully saturated, attention-getting hues on the right.

When selecting your palettes of colors for encoding categorical subdivisions in graphs, maintain two versions: one with soft, yet distinct hues, and another with bright versions of the same hues. This will allow you to choose colors from the subtle palette for items that don't need to stand out and choose colors from the bright palette for those that should. Here's an example of two versions of the 12 distinct hues that were listed previously, one for ordinary use and one for highlighting:

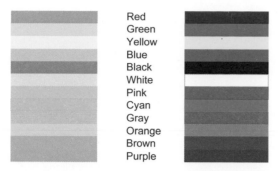

Red
Green
Yellow
Blue
Black
White
Pink
Cyan
Gray
Orange
Brown
Purple

FIGURE 6.19 This figure shows two standard palettes of distinct hues: one that includes only soft hues and one that includes only bright hues.

Another consideration when using hues to encode data is that about 10% of males and 1% of females suffer from color blindness, unable to distinguish certain hues because of a lack of cones (i.e., color receptors) in the eye that detect a certain range of hue. Most people who are color-blind lack the cones that enable them to distinguish between red and green, which both appear brown. Given this fact, especially when your audience is large and is likely to include people with color blindness of this type, it is best to use only red or green, but not both, unless you are careful to vary their intensities enough to make them appear as distinct shades of brown.

Limits to the Use of Contrast

Our visual perception evolved in a way that makes us particularly aware of differences, features that stand out in contrast to the rest. When we look at something, our eyes are not only drawn to differences, but our brains insist on making some sense of those differences. If you create a table of information and all the text is black except for one column that is red, readers will assume that the difference is significant and that the red text is especially important. It is up to us to only include visual differences that correspond to an actual differences in meaning.

Contrast is most effective when only one thing is different in a context of other things that are the same. As the number of differences increases, the degree to which those differences stand out decreases. In the graph below, the attributes were carefully controlled so that the market share for Our Company easily stood out from the rest because it is visually different from all of the other information.

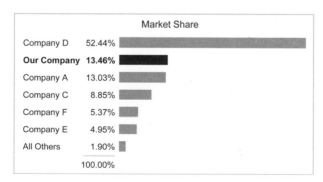

FIGURE 6.20 This graph uses contrast effectively to highlight the most important information.

If you use contrast to make so many things stand out that differences become the rule rather than the exception, your message will become lost in clutter, because you're trying to say too much at once. At that point it's time to break your complex message into multiple messages.

Gestalt Principles of Visual Perception

Many fields of scientific study have contributed to our understanding of visual perception, but none have revealed more of relevance to visual design than the Gestalt School of Psychology. It was the original intent of this effort, when it began in 1912, to uncover how we perceive pattern, form, and organization in what we see. The German word *gestalt*, however lofty it may seem to speakers of English, simply means "pattern." The founders observed that we organize what we see in particular ways in an effort to make sense of it. The result of their research was a series of Gestalt principles of perception, which are still respected today as accurate descriptions of visual behavior. These principles reveal visual attributes that incline us to group the objects that we see in particular ways. Many of these findings are relevant to our interest in the design of tables and graphs as well as to the overall design of the reports that contain them. The following is an introduction to these principles.

Principle of Proximity

We perceive objects that are close to each other as belonging to a group. In the following illustration, we naturally perceive the 10 circles as three groups because of the way they are spatially arranged:

FIGURE 6.21 This figure illustrates the Gestalt principle of proximity.

This principle applies very directly to the design of tables. We can direct our readers to scan predominantly across rows or down columns of data by spacing the data to accentuate either the row or the column groupings, as illustrated below:

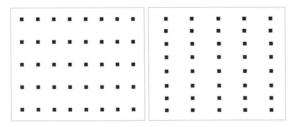

FIGURE 6.22 This figure illustrates an application of the Gestalt principle of proximity to table design.

In response to subtle grouping cues, we are naturally inclined to scan the objects in the left-hand box by row and those in the right-hand box by column. This is because in the left-hand box the objects are slightly closer together horizontally, which groups them by rows, and in the right-hand box they are slightly closer together vertically, which groups them by columns. The conscious use of space to group objects together or keep them separate is a powerful tool for organizing information and directing our readers to view it in a particular way.

Principle of Similarity

We tend to group together objects that are similar in color, size, shape, and orientation. Here are some examples:

FIGURE 6.23 These examples Illustrate the Gestalt principle of similarity.

This principle reinforces what we've already learned about the usefulness of color (both hue and intensity), size, shape, and orientation to encode categorical variables. If these attributes are used to encode a distinction between different data sets in a graph, they work quite effectively as long as the number of distinctions is kept to a reasonable minimum, and the distinctions are significant enough for us to clearly differentiate the groups. We can also use this principle to direct readers of tables to focus primarily across rows or down columns, as in the following example:

XXXXX XXXXX XXXXX XXXXX XXXXX XXXXX XXXXX XXXXX
XXXXX XXXXX XXXXX XXXXX XXXXX XXXXX XXXXX XXXXX
XXXXX XXXXX XXXXX XXXXX XXXXX XXXXX XXXXX XXXXX
XXXXX XXXXX XXXXX XXXXX XXXXX XXXXX XXXXX XXXXX
XXXXX XXXXX XXXXX XXXXX XXXXX XXXXX XXXXX XXXXX
XXXXX XXXXX XXXXX XXXXX XXXXX XXXXX XXXXX XXXXX
XXXXX XXXXX XXXXX XXXXX XXXXX XXXXX XXXXX XXXXX
XXXXX XXXXX XXXXX XXXXX XXXXX XXXXX XXXXX XXXXX
XXXXX XXXXX XXXXX XXXXX XXXXX XXXXX XXXXX XXXXX

FIGURE 6.24 This figure demonstrates the use of the principle of similarity to arrange data columnwise in a table.

Principle of Enclosure

We perceive objects as belonging together when they are enclosed by anything (e.g., a line or a common field of color) that forms a visual border around them. They appear to be set apart in a region that is distinct from the rest of what we see. Notice how strongly your eyes are induced to group the enclosed objects below:

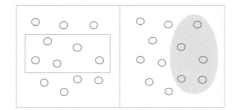

FIGURE 6.25 These two examples illustrate the Gestalt principle of enclosure; in the left-hand example, the enclosure is formed by a line, and, in the right-hand example, by a field of gray in contrast to the white that surrounds it.

The arrangement of the two sets of circles on the previous page is exactly the same, yet the differing enclosures lead us to group the circles in very different ways. This principle is exhibited frequently in the use of borders and fill colors in tables and graphs to group information and set it apart. As you can see in the previous illustration, it does not take a strong enclosure (e.g., bright, thick lines or dominant colors) to create a strong perception of grouping. In fact, we sometimes perceive enclosures that aren't really there, such as the white shape that you see in the middle of the black objects below:

FIGURE 6.26 This is an example of an *illusory contour* (i.e., perception of a white oval in the middle of the black objects, forming a separate region of white space).

The reason that we perceive this illusory enclosure is explained by the next two Gestalt principles: the *principle of closure* and the *principle of continuity*.

Principle of Closure

We have a keen dislike for loose ends. When faced with ambiguous visual stimuli—objects that could be perceived either as open, incomplete, and unusual forms, or as closed, whole, and regular forms—we naturally perceive them as the latter. In the example of an illusory enclosure above, it is natural to perceive the black objects as rectangles, circles, ovals, and continuing lines, positioned partially behind a white oval, rather than as a collection of oddly shaped objects. The principle of closure asserts that we perceive open structures as closed, complete, and regular (e.g., the black objects above) whenever there is a way that we can reasonably interpret them as such. Here's a more direct illustration of this principle:

FIGURE 6.27 This figure illustrates the principle of closure.

It is natural for us to perceive the lines on the left as a rectangle rather than as two sets of three connected and perpendicular lines and to perceive the object on the right as an oval rather than a curved line.

We can put this tendency to perceive whole structures to use when designing tables and graphs. For example, we can group objects (e.g., points, lines, or bars on a graph) into visual regions without the use of complete borders to define the space. It is not only sufficient but preferable to define the area of a graph through the use of a single set of subtle X and Y axes rather than by heavy lines and fill colors that enclose it completely, as in the following examples:

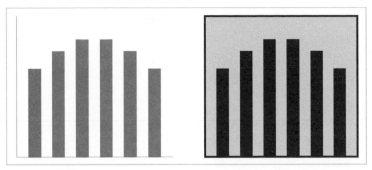

FIGURE 6.28 These examples demonstrate an application of the principle of closure. We can define the area of a graph with subtle X and Y axes, as seen on the left, thus minimizing the visual weight of a graph's supporting components in relation to its data components.

Principle of Continuity

We perceive objects as belonging together, as part of a single whole, if they are aligned with one another or appear to form a continuation of one another. In the left-hand illustration below, we see the various visual objects as forming a simple image of a rectangle and a wavy line. If we separated the two objects, we assume they would look like those in the middle illustration. We don't, however, see a rectangle and three curved lines, like those in the right-hand illustration, despite the fact that this is another possible interpretation of the image. Our tendency is to see the shapes as continuous to the greatest degree possible,

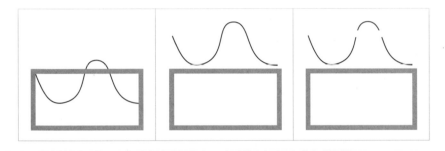

FIGURE 6.29 These examples illustrate the principle of continuity.

In the bar graph below, the left alignment of the bars makes it obvious that they share the same baseline, thus eliminating the need to reinforce the fact with a vertical axis:

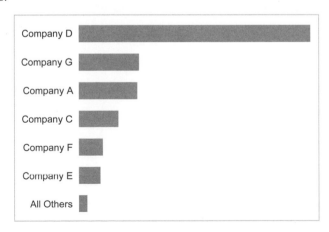

FIGURE 6.30 This is an example of an application of the principle of continuity to graph design.

In the table below, it is obvious which categorical subdivisions are division names and which are department names, based on their distinct alignment:

Division	Department	Headcount
G&A	Finance	15
	Purchasing	5
	Information Systems	17
Sales	Field Sales	47
	Sales Operations	10
Engineering	Product Development	22
	Product Marketing	5

FIGURE 6.31 This is an example of an application of the principle of continuity to table design.

The left alignment of the divisions and departments and the right alignment of the headcounts create a strong sense of grouping without any need for vertical grid lines to delineate the categories. Even if two columns of data overlap, with no white space in between, as long as they are separately aligned, they remain clearly distinguishable, as in the following example:

Division/Department	Headcount
G&A	
Finance	15
Purchasing	5
Information Systems	17
Sales	
Field Sales	47
Sales Operations	10
Engineering	
Product Development	22
Product Marketing	5

FIGURE 6.32 This is an example of an application of the principle of continuity to table design to save horizontal space.

In this example, the principle of continuity has been used to align related categories in a way that saves horizontal space.

Principle of Connection

We perceive objects that are connected (e.g., by a line) as part of the same group. In the example below, even though the circles are equally spaced from one another, the lines that connect them create a clear perception of two horizontally attached pairs:

Technically, the principle of connection was not directly presented by the original Gestalt School of Psychology but has been articulated since as an extension of the *principle of continuity*.

FIGURE 6.33 This figure illustrates the principle of connection.

Connection exercises even greater power over visual perception than proximity or similarity (color, size, and shape) but less than enclosure, as you can see from the examples below:

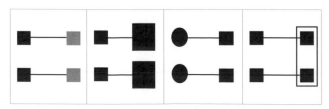

FIGURE 6.34 These are examples of the relative perceptual strength of the Gestalt principles that we've examined so far.

The perceptual strength of connection makes lines a useful method of encoding data in graphs. Remember our observation earlier in Chapter 5, *Fundamental Variations of Graphs*, that points without lines in graphs are much harder for our eyes to connect. Notice the difference between the two examples below:

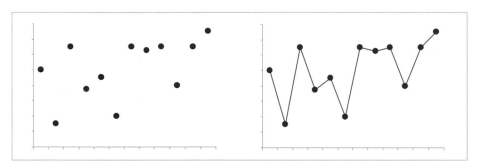

FIGURE 6.35 These examples illustrate the power of lines in graphs to reveal connectedness.

Lines in a graph not only create a clear sense of connection but also bring to light the overall shape of the data, which couldn't be discerned without them.

Summary at a Glance

Mechanics of Sight

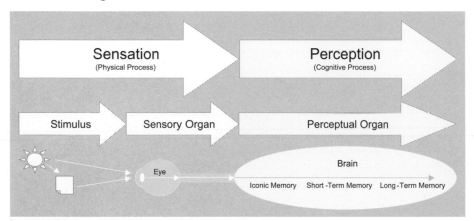

Attributes of Preattentive Processing

Attribute	Quantitative
Line length	Yes
2-D position	Yes
Orientation	No
Line width	Somewhat
Size	Somewhat
Shape	No
Curvature	No
Added marks	No
Enclosure	No
Hue	No
Intensity	Somewhat

Gestalt Principles of Visual Perception

Principle	Description
Proximity	Objects that are close together are perceived as a group.
Similarity	Objects that share similar attributes (e.g., color or shape) are perceived as a group.
Enclosure	Objects that appear to have a boundary around them (e.g., formed by a line or area of common color) are perceived as a group.
Closure	Open structures are perceived as closed, complete, and regular whenever there is a way that they can be reasonably interpreted as such.
Continuity	Objects that are aligned together or appear to be a continuation of one another are perceived as a group.
Connection	Objects that are connected (e.g., by a line) are perceived as a group.

7 GENERAL DESIGN
FOR COMMUNICATION

Based on an understanding of visual perception, you can build a set of visual design principles, beginning with those that apply equally to tables and graphs. The primary objective of visual design is to present content to your readers in a manner that highlights what's important, arranges it for clarity, and leads them through it in the sequence that tells the story best.

Highlight the data

Organize the data

Integrate tables, graphs, and text

Visual design can serve many purposes, not least of which is to create beauty, which can be experienced purely for its own sake. This is the work of the artist. Without it our lives would be flat and our souls malnourished. Artists spend their lives learning from the masters and their own painstaking experience. Through each stroke of the brush, angle of the chisel, or subtle positioning of the light, they generate beauty. As creators of tables and graphs, our use of visual design serves a different purpose but one that is also fundamental to life and deserves no less attention. We use visual design to *communicate*. There is a message in the numbers, which will be perceived and acted upon or will go unnoticed and ignored, depending on our knowledge of visual design and our ability to apply that knowledge to the important task of communication.

In this chapter we'll examine the aspects of visual design that apply equally to all visual forms for communicating quantitative information, including tables, graphs, and text. The general practices of communication-oriented design all support two fundamental objectives:

1. Highlight the data
2. Organize the data

We highlight the data to give them a voice that comes through loudly and clearly, without distraction. We organize data to lead readers through the contents in a manner that promotes optimal understanding and use of the information.

Highlight the Data

It is appropriate to begin this section by repeating six incisive words written by Edward Tufte: "Above all else show the data."[1] These words should serve as your mantra. Nothing is more central to your task.

1. Edward R. Tufte (1983) *The Visual Display of Quantitative Information.* Cheshire CT: Graphics Press, page 92.

Tufte introduced a useful concept that he calls the *data-ink ratio*. Here it is in his own words:

> *A large share of ink on a graphic should represent data-information, the ink changing as the data change. Data-ink is the non-erasable core of a graphic, the non-redundant ink arranged in response to variation in the numbers represented. Then,*
>
>> *data-ink ratio = data-ink / total ink used to print the graphic = 1.0 [minus the] proportion of a graphic that can be erased without loss of data-information . . .*
>
> *Maximize the data-ink ratio, within reason. Every bit of ink on a graphic requires a reason. And nearly always that reason should be that the ink presents new information.*[2]

The data-ink ratio concept applies equally to graphs and tables. If the total amount of ink used in a table or graph equals a value of one, what is the value of the portion of the ink that displays quantitative and categorical information versus all else (e.g., supporting components like grid lines or superfluous components like ornamentation that play no role whatsoever in supporting the data)? The greater the ratio of the ink that you use to communicate the data to the total amount of ink in the table or graph (i.e., the closer its value is to one), the better you've highlighted the data.

The object isn't to eliminate all *non-data ink*. To some degree you will always need supporting visual components to make tables and graphs readable. The object is to reduce the non-data ink to no more than what is necessary to make the data ink understandable.

We highlight data through a design process that consists of two types of activity:

1. Reducing the non-data ink
2. Enhancing the data ink

Reduce the Non-Data Ink

The process of reducing the non-data ink involves two steps:

1. Subtract unnecessary non-data ink
2. De-emphasize and regularize the remaining non-data ink

SUBTRACT UNNECESSARY NON-DATA INK

The process of subtracting unnecessary non-data ink involves asking the following question about each visual component: "Would the data suffer any loss of meaning or impact if this were eliminated?" If the answer is "no," then get rid of it. Resist the temptation to keep things just because they're cute or because you worked so hard to create them. If they don't support the message, they don't serve the purpose of communication. As the author Antoine de Saint-Exupery suggests: "In anything at all, perfection is finally attained not when there is no longer anything to add, but when there is no longer anything to take away."[3]

By subtracting what is not required to support the message, we bring our communication one step closer to *elegance*. The word *elegance* comes originally

2. Edward R. Tufte (1983) *The Visual Display of Quantitative Information.* Cheshire CT: Graphics Press, pages 93 and 96.

3. This quotation of Antoine de Saint-Exupery, and the explanation of the term *elegance* were taken from Kevin Mullet and Darrel Sano (1995) *Designing Visual Interfaces.* Sun Microsystems, Inc., page 17.

from the Latin term *eligere*, which means to choose out or to select carefully. To achieve elegance in communication, we must carefully select the content that is essential to the message and trim all else away.

DE-EMPHASIZE AND REGULARIZE THE REMAINING NON-DATA INK

Once you've subtracted all the unnecessary non-data ink, you should push what remains far enough into the background to enable the data to stand out clearly in the foreground. This can be achieved by using muted visual attributes for non-data ink components.

Tables and graphs consist of three visual layers: 1) the *data* as the top or prominent layer, 2) *supporting components* (e.g., grid lines) as the middle layer, and 3) the *background* as the surface on which the data and supporting components reside. The background of printed material is the paper of a particular color on which the content is printed. The background of on-line content is the computer screen of a particular color on which the material is projected. Non-data ink, consistent with its supporting role, should stand out just enough from the background to serve its purpose but not so much that it draws attention. This can be achieved through the use of thin lines and soft, neutral colors (e.g., light gray). To do otherwise, giving the same visual weight to data ink and non-data ink, gives no visual cues to lead the reader's eyes to what's important. When everything stands out, nothing stands out.

Because our eyes are drawn to contrast, you can go one step further to reduce the visibility of non-data ink by making it as *consistent* as possible, so that none stands out. Multiple instances of the same supporting components throughout a report should look precisely the same everywhere they appear. Any differences work against your purpose by inviting your readers' eyes to notice and their brains to assign meaning to those differences.

Take a few minutes now to examine two or three of your own reports to identify opportunities to reduce the non-data ink. You may be surprised to find how much there is that could be subtracted, muted, and regularized for greater effect.

· · · · · · ·

Enhance the Data Ink

You can enhance the data ink through a process that consists of two steps:

1. Subtract unnecessary data ink
2. Emphasize the most important data ink

SUBTRACT UNNECESSARY DATA INK

You must carefully avoid saying too little by trying to say too much. Not all data are equally important. This is especially true when your readers don't have the time or the patience necessary to savor a message in all its subtlety. Don't remove anything that is important, but be sure to remove all that is peripheral to the interests and purposes of your readers. Every step of data reduction causes what remains to stand out even more. The more you earn your readers' trust by giving them only what they need, the more they'll pay attention to everything you do give them.

The intention here is to summarize when detail isn't necessary and to trim away what's not important, not to arbitrarily reduce the content of your message. It's appropriate for a single table or graph to deliver a great deal of information or to articulate a complex (but not overly complex) message. Aim to give your readers what they need, and all that they need, but nothing more.

EMPHASIZE THE MOST IMPORTANT DATA INK

Data values are encoded differently in tables than in graphs. In tables, they are encoded entirely in verbal language (i.e., words and numbers), but in graphs they are encoded primarily in visual language (e.g., points, lines, and bars) although words and numbers are used as well. Regardless of the encoding method, certain visual attributes of objects, words, and numbers stand out more than others.

In the previous chapter on visual perception, you learned that preattentive visual attributes differ in the degree to which they stand out. Size is a good example. You can make something stand out by making it bigger. Objects, words, and numbers that are bigger stand out more than those that are smaller, all else being equal.

You can take advantage of this to emphasize the most important data ink relative to the rest. Here is a list of those preattentive visual attributes that are especially useful for emphasizing data ink in tables and graphs:

Attribute	Values Useful for Emphasis
Line width	Thicker lines (including words and numbers that are boldfaced) stand out more than thinner lines.
Orientation	Slanted words and numbers (i.e., italics) stand out more than those that are oriented normally (i.e., not slanted), assuming that vertically oriented type is the norm.
Size	Bigger objects, words, and numbers stand out more than smaller objects.
Enclosure	Objects, words, and numbers that are enclosed by lines or background fill color stand out more than those that are not enclosed.
Hue	Objects, words, and numbers that have any hue that are different from the norm stand out.
Color intensity	Objects, words, and numbers that are bright stand out more than those that are light or pale.

Each step for highlighting the data results in greater *simplicity*. In the communication of quantitative information, simplicity of design is the essence of *elegance*. Your message may be complex, but its design—the form in which you present it—should be so simple that to your readers it is nearly invisible.

Organize the Data

When your readers look at a page or screen of information, they immediately begin to organize what they see in an effort to make sense of it. As a designer of communication, it is your job to organize the information for them in a manner that supports the clarity of the message. If you fail to do this effectively, the result is not information that is *unorganized*, but information that is organized in a manner that does not support its essential message, resulting in ineffective communication. In fact, your readers may get a different message entirely— perhaps one that is wrong. Communication involves much more than knowing what to say; it also involves knowing how to say it.

The page or screen that serves as your medium of communication will often contain more than a single table or graph. Your message may require multiple tables, multiple graphs, or a combination of both, along with additional text in the form of bullet points, sentences, or even whole paragraphs. When you arrange this information on the page or screen, you must do so consciously to tell a story. What should I say first? What should I save for last? What should I emphasize more than the rest? The answers to these questions take on the form of visual attributes designed to accomplish the following:

1. Group the data (i.e., segment the data into meaningful subsets)
2. Prioritize the data (i.e., rank the data by importance)
3. Sequence the data (i.e., give direction for the order in which the data should be read)

Group the Data

You must always begin with a clear sense of what belongs together, what your readers should perceive as belonging to the same group because those units of information have something in common. Once this is clear, you can select from various visual design techniques that can be used to organize the information into groups.

Grouping takes place on several levels. You begin with your overall quantitative message and then break down its content into different topics. The various topics are then grouped into the appropriate modes of expression: tables, graphs, and text. Within tables and graphs, data are naturally grouped into categorical and quantitative information. Finally, categorical information is grouped into its various subdivisions, and quantitative information is grouped as quantitative values associated with each of those categorical subdivisions.

It's your job to make this grouping obvious to your readers. It shouldn't be up to your readers to do the work of arranging the content into meaningful groups when you can do this in advance for them.

The Gestalt principles of visual perception reveal a number of techniques that can be used to group data meaningfully. The simplest approach—*proximity*—is often the best. This is especially true for arranging content into various topics. If you were communicating quarter-to-date sales performance to a sales manager, your overall message might consist of regional sales performance compared to forecast, the top orders, and the top customers. You would have a single message consisting of three related topics. By placing the information related to each topic close together and by separating the topic groups by white space, you create a simple and clear arrangement with nothing to hinder your readers' eyes as they move from one group to another.

Sometimes your message consists of several separate but related topics that need to be appropriately grouped and arranged on a page or screen, and among those individual topics reside relationships that should be identified. Let's continue the previous example. The quarter-to-date sales performance information consists of three primary topics, but each includes sales expressed both as bookings and billings. Let's assume that you have appropriately arranged the three topics into a graph that displays regional sales performance compared to forecast, a table listing the top orders, and a table listing the top customers, each separated by enough white space to render it distinct. Even though bookings and billings both appear in the graph as well as the tables, it would be helpful to tie the separate instances of each together visually. This would make it easier for your readers to quickly scan for all bookings information separate from billings information, and vice versa. You could do so simply by selecting one of the remaining Gestalt principles or one of the preattentive attributes that you learned about in the last chapter on visual perception. Given this scenario, what attribute or principle might you choose to visually group bookings as distinct from billings? Take a moment to run through the list of available methods, weighing the potential advantages of each.

· · · · · · ·

There is certainly no one right answer. You may have recognized, though, while assessing the alternatives, that many of the available methods suggest that either bookings or billings is more important than the other. For instance, if you chose color intensity to distinguish bookings from billings, rendering bookings as black and billings as gray, bookings would stand out perceptually as more prominent. This would be appropriate if the purpose of your message were to emphasize bookings over billings, but if you wished to treat them equally, color intensity would not be the best method. You could, however, select different hues for each. As long as you avoided hues like black and red, which tend to be perceived as visually prominent in contrast to other hues, this choice would do the job effectively.

Tables and graphs both use conventional means to organize data into categories. Tables primarily use the Gestalt principles of *proximity* and *continuity* to organize different categories into columns and rows. Graphs use many techniques, such as the principles of *similarity* (e.g., common hues or shapes) and *connection* (e.g., the use of a line to connect points) to group data. We'll examine these techniques in the chapters on table and graph design.

Prioritize the Data

Whenever you communicate quantitative information, it is imperative that you step back from it and ask yourself, "What are the important numbers here?" Once you've established a clear sense of what's important, you should make that information stand out clearly from the rest. This is a vital part of your job. Don't just highlight important numbers when you happen to think about it or when their importance hits you over the head. Think about it every time.

People who use numbers routinely in their jobs often don't have an effective strategy or even any strategy at all for approaching the numbers. Not all numbers are equally important. In fact, some numbers are so much more important than others that a few seconds spent examining and understanding them produces benefits that could never be equaled by years of concentration on all the others. Help your readers develop a productive approach to numbers by pointing out the ones that are most deserve their attention.

Some visual attributes are perceived quantitatively. Their values can be arranged along a continuum ranging from less to more, small to big. Such attributes have the built-in ability to make some information stand out as more prominent than the rest. Here's a list of the visual attributes that are perceived quantitatively, along with examples of how each can be used to highlight important sections of text, tables, and graphs:

Attribute	Tables and Text	Graphs and Objects
Line width	• Boldfaced text	• Thicker graph lines
Size	• Bigger tables	• Bigger graphs
	• Larger fonts	• Wider bars
		• Bigger symbol shapes
Color intensity	• Darker or brighter colors	• Darker or brighter colors
2-D position	• Positioned at the top	• Positioned at the top
	• Positioned at the left	• Positioned at the left
	• Positioned in the center	• Positioned in the center

As the last item in the list suggests, certain positions in a 2-D space stand out as more prominent than the rest. Because many written languages are read from left to right and top to bottom, you can make text, tables, or graphs appear more prominent by locating them at the top left of a page or screen. Secondarily, anything located anywhere along the left edge of the space also appears prominent, though slightly less so. Lastly, anything located in the center of the space, if visually set apart in some way from the content around it, also grabs attention, all else being equal.

In addition to these quantitatively perceived visual attributes, you can also take advantage of the fact that visual *contrast* of any type makes the contrasting information stand out from the norm. Here are additional attributes that you can use to make particular information stand out as important by means of contrast:

Attribute	Tables and Text	Graphs and Objects
Line orientation	• Italics	• Data points with an orientation that is different from the norm
Shape	• Any font that is different from the norm	• Any symbol shape that is different from the norm
Enclosure	• Border around table, rows, columns, or particular values	• Border around graph or particular values
	• Fill color behind tables, rows, columns, or particular values	• Fill color in a graph or behind particular values
Additional marks	• Underlined text	• Symbol shapes that have an additional mark (e.g., a line through a circle when the other symbols are simple circles)
Hue	• Almost any hue that is different from the norm	• Almost any hue that is different from the norm
2-D position	• Any position that is out of vertical or horizontal alignment with the norm	• Any position that is out of vertical or horizontal alignment with the norm

The last attribute of *2-D position* highlights the significance of alignment in visual design. We are more sensitive to the vertical and horizontal alignment of text and objects than you might imagine. The slightest misalignment jumps out at us, and we react by trying to impose meaning on that contrast. Unless you intend to make something stand out as different, be careful to keep the edges of text and objects aligned so that your readers' eyes can scan down and across without disruption.

You may use differences in horizontal alignment quite consciously to establish a hierarchical relationship between different sections of content, with subordinate content indented to the right of major content. When you use indentation in this manner, be sure to indent far enough to make your intention clear. Using alignment in this manner makes it easy for your readers to separately scan the major content without distraction from the subordinate content when they wish to take in the main points quickly.

There is actually one more method that we haven't covered because it doesn't involve a visual attribute of objects but is instead a functional type of object. I'm referring to a collection of objects called *pointers*. This includes objects like arrows, asterisks, and check marks. Put one of these next to or pointing to any content, and your reader's attention will definitely be drawn to it. This works due to the preattentive perception of *added marks* combined with the conventional significance of a pointer as a sign of importance. Pointers are not subtle, especially arrows, so you should use them with discretion to avoid visual clutter.

Sequence the Data

The final objective of visual design is to provide clear directions for your readers regarding the best sequence for navigating through a report's contents. The strongest sequencing method is the location of content in 2-D space. Because we read from left to right and top to bottom, this is generally the order in which your readers will scan the page or screen. If you clearly divide the contents into columns, such as those in newspapers, readers will first scan the left-most column from top-to-bottom, then move to the top of the next column.

The strength of this left-to-right and top-to-bottom navigational sequence is greatest with textual content because text can only be perceived through the sequential process of reading. This same sequence works for non-text objects as well (e.g., graphs) but not as strongly. For instance, if your page contains a collection of graphs without sections of text to introduce them, readers will still give attention to each graph in the normal left-to-right, top-to-bottom sequence, all else being equal. However, if any one of the graphs has been highlighted as important using any of the prioritizing methods noted in the previous section, readers' eyes will likely be drawn immediately to that graph. If your message requires that your readers work their way through a collection of tables and/or graphs in a particular order, you can further reinforce the navigational route by using numbers (1, 2, 3, etc.), alphabetical letters (A, B, C, etc.), or some other form of sequential labeling.

Integrate Tables, Graphs, and Text

Tables, graphs, and text form a powerful team, working together intimately to communicate quantitative information. Each brings a different set of strengths to the task of communication. We've already examined the separate strengths of tables and graphs. In this section, we'll focus on the contribution of text and the way it can be integrated with tables and graphs to create clear and powerful messages.

The Role of Text

To complement or enhance tables and graphs, text can:

- Label
- Introduce
- Explain
- Reinforce
- Highlight
- Sequence
- Recommend
- Inquire

Let's take a quick look at each role.

LABEL
We've already examined the role of labeling. Tables and graphs both use text to label data. Tables use text (i.e., words and numbers) not only to express quantitative and categorical data but also to label what the columns and rows contain. Graphs incorporate text in the form of titles, labels for categorical subdivisions and quantitative values along the scale lines, and legends to interpret the visual

encoding of categorical subdivisions (e.g., the blue bars represent the eastern sales region). Text in the form of labels supplies critical information to enable readers to interpret tables and graphs.

Reports containing tables and graphs also use text in the form of titles. Clear titles are vital data in themselves. How many times have you seen a report with a title that revealed nothing definitive about its contents? When people scan lists of available reports in an effort to find one that contains the information they need, they often do this with no information other than the titles. Good titles are invaluable.

INTRODUCE

Quantitative displays often require an introduction to set the reader on a clear path to understanding. Text is the ideal medium for providing introductions.

Introductions are especially useful in new reports and for new readers of old reports, potentially saving readers a great deal of time and frustration. Among other things, an introduction should preview what the readers will find in the report, what they should especially notice, and what they should do with the information. Because you can't always hand a report directly to all its eventual readers, the introduction is your chance to set the stage for the report using text that states what you would tell them in person if you could.

EXPLAIN

An introduction to a report is not the best place to put every bit of text that may be needed to explain the data it contains. Explanations work most effectively when they are provided right at the point where they're needed to clarify some-thing about the message. If you provide a time-series graph that displays an unusual brief up-tick in sales during the month of May, you may want to men-tion right there, in or just underneath the graph, that a successful marketing campaign beginning in late April was responsible for the anomaly. If a few words are what it takes to make the message clear, then they belong there. Whenever a table or graph doesn't speak clearly enough on its own, its design should be improved, or a little text should be added.

REINFORCE

Some information is so important that you should say it more than once and in more than one way to increase its likelihood of getting through to your readers. By encoding that information visually in a graph or verbally in the columns and rows of a table, and then doing so again with a few well-chosen words, you will use text to increase the odds that the message will be heard. We tend to better remember what has been presented to us in both pictures and words. You don't want to overdo it though. Don't say everything in multiple ways or you'll waste your readers' time and lose their confidence. The important stuff, however, deserves a little extra.

HIGHLIGHT

We examined several methods for visually highlighting important data. Some-times it's also useful to highlight particular information by referencing it with words as well. This is different from reinforcement because in this case you're not repeating the information in a different form; you're simply calling the reader's attention to it. For instance, if the sales ranking of a particular product warrants special notice, you may say so in words right in or underneath the graph. Perhaps it isn't appropriate to make that product stand out above the rest visually in the graph itself because that would distract from the other products that are also important, but a short note following the graph could do the job without creating a visual distraction in the graph.

SEQUENCE

Sometimes it's challenging to use visual methods alone to clearly reveal the order in which your readers should examine the contents of a report. Perhaps information in a report cannot be positioned from left to right and top to bottom in the order it should be read because of a greater need to use 2-D location to highlight the importance of some data or to group data in a particular way. In circumstances such as these, you can use text to instruct your readers to navigate through the contents in a particular way.

RECOMMEND

As a communicator of important quantitative information, your job often involves more than simply informing. Sometimes it's your responsibility to recommend what could or should be done. Recommendations for action are best communicated in words. Whether or not this is your explicit role, taking the initiative to make recommendations that you deem warranted may be greatly appreciated.

INQUIRE

Inquiry is vastly underrated and too often ignored. Quantitative information often invites questions that ought to be asked. You can sometimes add more value to your business by asking a single important question than by providing hundreds of answers. I'm a bit biased in favor of thoughtful inquiry. We so often get caught up in *business as usual* that we fail to question why things are as they are or whether things might be better if they were different. As a communicator of quantitative business information, you're in a great position to recognize opportunities for further exploration, further speculation, and valuable questions that somebody ought to be considering. Why not ask such questions by placing a few words in your reports near the information that prompted them? I realize that your readers may respond by assigning you the task of exploring those questions, but who better? If you're like me, discovering the right questions to ask and then doing the research and analysis to find the right answers is the real fun of working with numbers.

Text Placement

The right place to position text, no matter what role it plays in the communication of quantitative information, is as close as possible to where it is needed. There are no rules that say you can't put text right in the middle of a graph or right next to a table with an arrow pointing to a particular number, row, or column. Tables, graphs, and text are complementary. There is no need to arbitrarily relegate them to separate areas in a report. Blend them together, placing each unit of content precisely where it is most needed. Just be careful when placing text in the plot area of a graph to do so in a way that does not obscure the shape of the data.

The reason for this practice is rooted in what you learned about visual perception in the last chapter. Our eyes have a limited area on which they can focus at any one time. If you place the legend for a graph too far from the data it interprets, you force your readers to jump back and forth over and over to read the graph because they can't keep all the encodings (e.g., the blue line represents widgets) in short-term memory. If you place the explanation for a table that appears on page 1 at the end of the report on page 10, you're causing unnecessary effort and frustration.

Perhaps your message involves a great deal of text spread across several pages, which refers to a single table or graph. In this case, you may actually want to reproduce the graph in multiple locations in order to always include it where it's needed.

You may have noticed that in this book I don't follow the traditional practice of placing notes and references either at the bottom of the page, the end of the chapter, or worst of all, the end of the book. I also don't force you to turn to a middle section of illustrations but instead have integrated all illustrations right where they're needed. This was a conscious design choice to facilitate communication. You face similar design choices regarding the integration of tables, graphs, and text whenever you construct a quantitative message. The tighter the integration, the better.

Required Text

Text should be included on every page of every report to answer the following questions:

- What?
- When?
- Who?
- Where?

Excerpts from multi-page reports are often copied and distributed. If the information that identifies the report's contents only appears at the beginning, readers will have no way of knowing what they're seeing when they have only a portion of the report. It takes only a minute to include this identifying information in the page header or footer of your reports.

WHAT

A good title is invaluable. A simple glance at the title should clearly tell your readers what the report contains. It should describe, without getting too long winded, the type of quantitative and categorical information that the report presents. The title *Sales* isn't enough. How is this report different from all the other reports that deal with sales? A title like *2002 Bookings by Month and Region* says a great deal more.

WHEN

Two facts should be provided with every report to inform your readers about its relation to time:

- The range of dates the information represents
- The point in time when the information was collected

Does the information represent a single hour, day, week, month, quarter, year? Does it represent some range of hours, days, etc.? Perhaps it represents an odd range of time, such as from April 23rd of 2002 through January 14th of 2003. Whatever the range, if it isn't clearly labeled in the table or graph, then make sure it is in the title or subtitle.

The point in time when the information was collected is often called the *as-of date*. "This represents expenses for February as of March 4th." This information is important because more expenses could be recorded later, or corrections could still be made to expenses after March 4th. Multiple reports covering the same period of time often differ simply because they were produced at different times, and the data changed in between them. A simple *as-of* followed by the date when the information was collected, noted in the header or footer of the report, conveniently satisfies this need.

WHO

The purpose of including your name or the name of the group you represent on your reports is not self-promotion; it is to let people know whom to contact if they have questions. I've spent many frustrating hours during the course of my career trying to track down the creator of a report because I needed to ask a simple question about it. Save your readers this annoyance. Provide your name, along with some means to contact you, such as an email address or phone number.

WHERE

By *where* I am referring to page numbers, which tell your readers where they are in a multi-page report. Try describing to someone where he or she can find a particular piece of information in a multi-page report that doesn't include page numbers. I find that the format *Page # of ##* (e.g., *Page 13 of 197*) is best because it informs your readers from the very first page how many pages they're facing in total. This is especially helpful when reports are distributed and read electronically because there is no physical stack of pages to alert readers to the size of the report. Have you ever started to print an electronic report only to realize later when you saw the line of angry coworkers at the printer that it was more than a thousand pages long?

Summary at a Glance

General Design Objectives of Quantitative Communication

HIGHLIGHT THE DATA

- Reduce the non-data ink
  - Subtract unnecessary non-data ink
  - De-emphasize and regularize the remaining non-data ink
- Enhance the data ink
  - Subtract unnecessary data ink
  - Emphasize the remaining data ink

ORGANIZE THE DATA

- Group the data
- Prioritize the data
- Sequence the data

Highlight What's Important

USING QUANTITATIVELY PERCEIVED VISUAL ATTRIBUTES

Attribute	*Tables and Text*	*Graphs and Objects*
Line width	• Boldfaced text	• Thicker graph lines
Size	• Bigger tables • Larger fonts	• Bigger graphs • Wider bars • Bigger symbol shapes
Color intensity	• Brighter, more vivid colors	• Brighter, more vivid colors
2-D position	• Positioned at the top, left, or center	• Positioned at the top, left, or center

USING VISUAL ATTRIBUTES IN CONTRAST TO THE NORM

Attribute	*Tables and Text*	*Graphs and Objects*
Line orientation	• Italics	• Data points with an orientation that is different from the norm
Shape	• Any font that is different from the norm	• Any symbol shape that is different from the norm
Enclosure	• Border around or shading behind table, rows, columns, or particular values	• Border around or shading behind graph or particular values
Additional marks	• Underlined text	• Symbol shapes that have an additional mark (e.g., a line through a circle when the other symbols are simple circles)
Hue	• Almost any hue that is different from the norm, especially black and red	• Almost any hue that is different from the norm, especially black and red
2-D position	• Any position that is out of vertical or horizontal alignment with the norm	• Any position that is out of vertical or horizontal alignment with the norm

Sequence the Data

- Using left-to-right, top-to-bottom positioning
- Using visual highlighting
- Using sequential labels (e.g., 1, 2, 3 . . .)

Required Page Information

What it is	*When it is*	*Who produced it*	*Where readers are*
In the form of a good title	In the form of the range of dates and an "as of" date	So readers know whom to contact	In the form of page numbers

8 TABLE DESIGN

Once you've determined that a table should be used to communicate your message and the type of table that will work best, you must refine your design so that the table can be quickly and accurately understood and used by your readers.

Structural components of tables
    Data components
    Support components
    Visual attributes of components
    Table terminology
Best practices of table design
    Delineating columns and rows
    Arranging data
    Formatting text
    Summarizing values
    Giving page information

We use tables regularly to communicate lists of quantitative information. They have become commonplace since the advent of spreadsheet software in the 1970s. When used properly and designed well, they're fantastic. Unfortunately, they often fall far short of their potential. Fortunately, the design practices required to optimize their effectiveness are quite simple and easily learned.

Structural Components of Tables

The components that we combine to construct tables and graphs fall into two categories:

- Data components
- Support components

Each component, whether its role is to directly express data or to support data by arranging or highlighting them, must be thoughtfully designed to fulfill its role effectively. When these components are designed well, the result is clear and efficient communication.

Data Components

Tables encode data as text (i.e., words and numbers). They include both categorical subdivisions and quantitative values. Tables generally work best when they also contain additional text that is used to complement the categorical subdivisions and quantitative values in various ways. In Chapter 7, *General Design for Communication*, we noted that additional text can do the following:

- Label
- Introduce
- Explain
- Reinforce
- Highlight
- Sequence
- Recommend
- Inquire

Text that is included in tables to perform any one of these functions gives additional information that is pertinent and useful to the table's message. In tables, such text appears most commonly as *titles* and *headers*. Headers are used to label columns of data (i.e., column headers) or one or more rows of data (i.e., row or group headers).

In total, then, tables and graphs of quantitative information actually consist of three types of data:

- Categorical subdivisions
- Quantitative values
- Complementary text

Support Components

Support components are the *non-data ink* objects that highlight or organize the data. The following visual objects and attributes can function as support components:

- White space and page breaks
- Rules and grids
- Fill color

WHITE SPACE AND PAGE BREAKS

The defining structural characteristic of tables is the arrangement of data into columns and rows. The primary visual means that we can use to support this structure is *white space*. Technically, the term *white space* is a bit of a misnomer in that the space it defines is only white if the color of the background is white. The alternate term *blank space* is technically more accurate, but we'll stick with the conventional term to keep things simple.

White space is used to group data objects that belong together. It is the visual mechanism that underlies the Gestalt principle of proximity. If it were not for the white space that separates groups of data objects, those objects would not appear to be close together. White space is not the most visually powerful means of grouping objects (e.g., the Gestalt principle of enclosure is more powerful), but its subtlety makes it especially useful. Properly used, white space organizes data objects into groups quite clearly without drawing attention to itself. Any design method that works effectively without drawing attention away from the data is invaluable.

Page breaks are a logical extension of white space. They can be used to group data by starting each new group on a separate page or screen. Once again, the underlying principle is proximity.

RULES AND GRIDS

The objects that go by the names *grids* and *rules* are two variations on the same theme. Both use lines to delineate or highlight data. Grids are combinations of horizontal and vertical lines that intersect to form rectangles around data. Rules are lines that run horizontally or vertically but do not intersect. Grids work on the Gestalt principle of enclosure, and rules work primarily on the principle of connection even though they do not actually touch the data that they underline but merely come close.

FILL COLOR

2-D areas containing data in a table can be filled with different intensities of color (i.e., variations of saturation or lightness, such as shades of gray) or different hues to group areas of the table together or highlight them as important. Both use the Gestalt principles of similarity and enclosure to define an area within a table (e.g., one or more columns and/or rows) and make the area stand out.

The surface on which all the data and support components reside is called the *background*. The background provides a clear surface that allows the data, as well as the support components that arrange and highlight the data, to stand out for easy and efficient reading.

Visual Attributes of Components

All the components of tables (data and support) possess visual attributes. Design not only involves choosing the components that are used to construct the table and display the information, but also choosing the visual attributes of each. Throughout the rest of this chapter, we will explore the best choices for each situation.

Table Terminology

It is helpful, when discussing table design, to begin with a standard set of terms for the various parts of a table. *Figure 8.1* on the next page provides the terms that we'll use throughout this chapter.

Let's clarify a few of the terms that may be new or ambiguous. In the context of tables that contain quantitative information, the term *body* generally refers exclusively to the rectangular area that contains the quantitative values. Only the area that is shaded in light gray in *Figure 8.1* constitutes the body of this example table. Defining *body* in this manner highlights the centrality of the quantitative information, reinforcing the fact that the primary message is in the numbers.

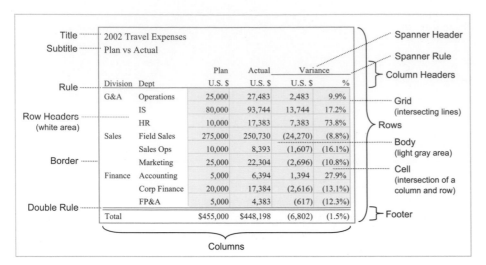

FIGURE 8.1 This diagram labels the parts of a table.

This figure is not meant to illustrate the best practices of table design but simply to label the parts.

Rows that summarize information contained in preceding rows are called *footers*. These can be used to summarize the entire table, as in the diagram above, or to summarize a subset of rows, which are called *group footers*.

The two left-most columns in the table above contain categorical subdivisions. When categorical subdivisions appear to the left of the quantitative values associated with them—a common arrangement—these subdivisions serve as labels for the rows and are therefore called *row headers*.

When a header spans multiple columns, it is called a *spanner header*. When a rule is used to underscore a spanner header and does so by spanning each of the columns to which the header applies, it is called a *spanner rule*.

Best Practices of Table Design

In this section we'll dig into the details of table design. We'll cover several topics in detail, grouped into five categories:

- Delineating columns and rows
  - White space
  - Rules and grids
  - Fill color
- Arranging data
  - Columns or rows
  - Groups and breaks
  - Column sequence
  - Data sequence
- Formatting text
  - Orientation
  - Alignment
  - Number and date format
  - Number and date precision
  - Fonts
  - Emphasis

- Summarizing values
  - Column summary values
  - Row summary values
  - Group summary values
  - Headers versus footers
- Page information
  - Repeated column headers
  - Repeated row headers

Delineating Columns and Rows

The design process involves several decisions regarding the layout of columns and rows to provide a structure that is easy and efficient to read and understand. Readers should be able to scan quickly through the content to find what they need and perhaps make localized comparisons of related numbers.

WHITE SPACE

White space is the preferred means for arranging data into columns and rows. The subtle use of blank space to group data into columns and rows is the least visible means available. Structural mechanisms of design that don't grab attention are the best when most possible.

The ability of white space to effectively delineate columns and rows is only curtailed when the overall space available for the table is so restricted that white space alone can't keep the columns or rows sufficiently distinct. If two rows of data are too close together, our eyes are not able to easily track across one without confusing it with the other. The same is true, but to a lesser extent, with columns.

As rows grow wider across, you need more white space between them to enable your readers' eyes to track across them without difficulty. When faced with wide rows, you have two potential means to address them using white space alone: 1) Decrease the horizontal white space between the columns to clarify the continuity of the rows, or 2) increase the vertical white space between the rows. Obviously, you can only decrease the horizontal white space between the columns so much before you reach the point where the columns are no longer clearly distinct. You can then add more white space between the rows, but too much can result in too little data on the page or screen. To preserve data density, you sometimes need to add a visual component other than white space (e.g., rules) to separate the rows. This balance between white space and overall data density is upset when the vertical white space between the rows exceeds the vertical space used by the rows of data themselves.

Let me illustrate. When I created *Figure 8.2* on the next page using *Microsoft Excel*, I selected a 10-point font (i.e., one that is 10 points high) and a row height of 12 points, resulting in 2 points of vertical white space between the rows (i.e., 20% of the height of the data).

Product	Jan	Feb	Mar	Apr	May	Jun	Jul	Aug	Sep	Oct	Nov	Dec
Product 01	93,993	84,773	88,833	95,838	93,874	83,994	84,759	92,738	93,728	93,972	93,772	99,837
Product 02	87,413	78,839	82,615	89,129	87,303	78,114	78,826	86,246	87,167	87,394	87,208	92,848
Product 03	90,036	81,204	85,093	91,803	89,922	80,458	81,191	88,834	89,782	90,016	89,824	95,634
Product 04	92,737	83,640	87,646	94,557	92,620	82,872	83,626	91,499	92,476	92,716	92,519	98,503
Product 05	86,245	77,785	81,511	87,938	86,136	77,071	77,773	85,094	86,002	86,226	86,043	91,608
Product 06	88,833	80,119	83,956	90,576	88,720	79,383	80,106	87,647	88,582	88,813	88,624	94,356
Product 07	82,614	74,511	78,079	84,236	82,510	73,826	74,498	81,511	82,382	82,596	82,420	87,751
Product 08	85,093	76,746	80,421	86,763	84,985	76,041	76,733	83,957	84,853	85,074	84,893	90,384
Product 09	87,646	79,048	82,834	89,366	87,535	78,322	79,035	86,475	87,399	87,626	87,440	93,095
Product 10	90,275	81,420	85,319	92,047	90,161	80,672	81,406	89,070	90,021	90,255	90,063	95,888

FIGURE 8.2 This example shows vertical white space between rows of data that is slightly less that optimal for horizontal tracking.

Given a full twelve months of data spread horizontally across the page, it is somewhat difficult to track across the rows with this spacing.

In the example below, the same information is displayed with white space between the rows equal to the height of the rows.

Product	Jan	Feb	Mar	Apr	May	Jun	Jul	Aug	Sep	Oct	Nov	Dec
Product 01	93,993	84,773	88,833	95,838	93,874	83,994	84,759	92,738	93,728	93,972	93,772	99,837
Product 02	87,413	78,839	82,615	89,129	87,303	78,114	78,826	86,246	87,167	87,394	87,208	92,848
Product 03	90,036	81,204	85,093	91,803	89,922	80,458	81,191	88,834	89,782	90,016	89,824	95,634
Product 04	92,737	83,640	87,646	94,557	92,620	82,872	83,626	91,499	92,476	92,716	92,519	98,503
Product 05	86,245	77,785	81,511	87,938	86,136	77,071	77,773	85,094	86,002	86,226	86,043	91,608
Product 06	88,833	80,119	83,956	90,576	88,720	79,383	80,106	87,647	88,582	88,813	88,624	94,356
Product 07	82,614	74,511	78,079	84,236	82,510	73,826	74,498	81,511	82,382	82,596	82,420	87,751
Product 08	85,093	76,746	80,421	86,763	84,985	76,041	76,733	83,957	84,853	85,074	84,893	90,384
Product 09	87,646	79,048	82,834	89,366	87,535	78,322	79,035	86,475	87,399	87,626	87,440	93,095
Product 10	90,275	81,420	85,319	92,047	90,161	80,672	81,406	89,070	90,021	90,255	90,063	95,888

FIGURE 8.3 This example shows vertical white space between rows of data that works effectively for horizontal tracking though it is perhaps slightly more than is needed.

This is the practical limit: a one-to-one ratio of data height to white space height. Notice that your eyes track across these rows with little difficulty. In fact, even a little less white space between the rows would still work effectively.

In the next example, I've exceeded the practical limits of vertical white space, with white space that equals 150% the height of the data. In this case you certainly don't have any difficulty tracking across the rows, but too few rows now fit on the page. The reduction in data density is now excessive.

Product	Jan	Feb	Mar	Apr	May	Jun	Jul	Aug	Sep	Oct	Nov	Dec
Product 01	93,993	84,773	88,833	95,838	93,874	83,994	84,759	92,738	93,728	93,972	93,772	99,837
Product 02	87,413	78,839	82,615	89,129	87,303	78,114	78,826	86,246	87,167	87,394	87,208	92,848
Product 03	90,036	81,204	85,093	91,803	89,922	80,458	81,191	88,834	89,782	90,016	89,824	95,634
Product 04	92,737	83,640	87,646	94,557	92,620	82,872	83,626	91,499	92,476	92,716	92,519	98,503
Product 05	86,245	77,785	81,511	87,938	86,136	77,071	77,773	85,094	86,002	86,226	86,043	91,608
Product 06	88,833	80,119	83,956	90,576	88,720	79,383	80,106	87,647	88,582	88,813	88,624	94,356
Product 07	82,614	74,511	78,079	84,236	82,510	73,826	74,498	81,511	82,382	82,596	82,420	87,751
Product 08	85,093	76,746	80,421	86,763	84,985	76,041	76,733	83,957	84,853	85,074	84,893	90,384
Product 09	87,646	79,048	82,834	89,366	87,535	78,322	79,035	86,475	87,399	87,626	87,440	93,095
Product 10	90,275	81,420	85,319	92,047	90,161	80,672	81,406	89,070	90,021	90,255	90,063	95,888

FIGURE 8.4 This example shows vertical white space between rows of data that exceeds the practical limit of a 1:1 ratio of text height to white space height.

White space can be intentionally manipulated to direct your readers' eyes to either scan predominantly across the columns or down the rows. If you wish to lead your readers to scan down the columns, rather than across the rows, make the white space between the columns more pronounced than the white space between the rows. To direct readers to scan predominantly across the rows, simply do the opposite.

If you don't have enough overall space on the page or screen to use white space alone to delineate the rows or columns, it's time to move on to another method.

RULES AND GRIDS

Rules and grids can be used to 1) delineate columns and rows, 2) group subsets of data, and 3) highlight subsets of data. Of these, the delineation of columns and rows is the least effective use of rules and grids though this use is unfortunately quite common. The problem with rules and grids is that they break up the data. As you scan across rows or down columns of data that are broken up by lines, the lines distract the eye, promoting a strong perception of individual cells through the Gestalt principle of enclosure rather than a seamless flow of information. Examples make this argument most vividly. The examples below start with strong grids and gradually decrease the use of grids and rules to a minimum. Judge for yourself which table is easiest to read.

Product	Jan	Feb	Mar	Apr	May	Jun
Product 01	93,993	84,773	88,833	95,838	93,874	83,994
Product 02	87,413	78,839	82,615	89,129	87,303	78,114
Product 03	90,036	81,204	85,093	91,803	89,922	80,458
Product 04	92,737	83,640	87,646	94,557	92,620	82,872
Product 05	83,733	75,520	79,137	85,377	83,627	74,826
Total	447,913	403,976	423,323	456,705	447,346	400,264

FIGURE 8.5 This table uses a thick grid to delineate columns and rows and to enclose the entire table.

Product	Jan	Feb	Mar	Apr	May	Jun
Product 01	93,993	84,773	88,833	95,838	93,874	83,994
Product 02	87,413	78,839	82,615	89,129	87,303	78,114
Product 03	90,036	81,204	85,093	91,803	89,922	80,458
Product 04	92,737	83,640	87,646	94,557	92,620	82,872
Product 05	86,245	77,785	81,511	87,938	86,136	77,071
Total	450,425	406,241	425,697	459,266	449,854	402,508

FIGURE 8.6 This table uses a thick grid to delineate the header and footer rows, thick vertical and thin horizontal grid lines to delineate the columns and rows of the body, and dark rules to enclose the entire table.

Product	Jan	Feb	Mar	Apr	May	Jun
Product 01	93,993	84,773	88,833	95,838	93,874	83,994
Product 02	87,413	78,839	82,615	89,129	87,303	78,114
Product 03	90,036	81,204	85,093	91,803	89,922	80,458
Product 04	92,737	83,640	87,646	94,557	92,620	82,872
Product 05	83,733	75,520	79,137	85,377	83,627	74,826
Total	447,913	403,976	423,323	456,705	447,346	400,264

FIGURE 8.7 This table uses a thin black grid to delineate columns and rows, thick horizontal rules to delineate the header and footer rows from the body, and thick black rules to enclose the entire table.

Product	Jan	Feb	Mar	Apr	May	Jun
Product 01	93,993	84,773	88,833	95,838	93,874	83,994
Product 02	87,413	78,839	82,615	89,129	87,303	78,114
Product 03	90,036	81,204	85,093	91,803	89,922	80,458
Product 04	92,737	83,640	87,646	94,557	92,620	82,872
Product 05	86,245	77,785	81,511	87,938	86,136	77,071
Total	450,425	406,241	425,697	459,266	449,854	402,508

FIGURE 8.8 This table uses a gray grid to delineate columns and rows, black horizontal rules to delineate the header and footer rows from the body, and black rules to enclose the entire table.

Product	Jan	Feb	Mar	Apr	May	Jun
Product 01	93,993	84,773	88,833	95,838	93,874	83,994
Product 02	87,413	78,839	82,615	89,129	87,303	78,114
Product 03	90,036	81,204	85,093	91,803	89,922	80,458
Product 04	92,737	83,640	87,646	94,557	92,620	82,872
Product 05	83,733	75,520	79,137	85,377	83,627	74,826
Total	447,913	403,976	423,323	456,705	447,346	400,264

FIGURE 8.9 This table uses gray rules to delineate the rows in the body and black rules to delineate the header and footer rows from the body.

Product	Jan	Feb	Mar	Apr	May	Jun
Product 01	93,993	84,773	88,833	95,838	93,874	83,994
Product 02	87,413	78,839	82,615	89,129	87,303	78,114
Product 03	90,036	81,204	85,093	91,803	89,922	80,458
Product 04	92,737	83,640	87,646	94,557	92,620	82,872
Product 05	83,733	75,520	79,137	85,377	83,627	74,826
Total	447,913	403,976	423,323	456,705	447,346	400,264

Figure 8.10 This table uses nothing but gray rules to delineate the header and footer rows from the body.

The last of these tables is the easiest to read. In it, the use of grids to delineate rows and columns has been entirely abandoned, and rules have only been used to separate the headers and footers from the body. Nothing more is needed. Anything more would reduce the table's effectiveness.

Even using rules to form a boundary around the entire table is a waste of ink although there is an exception, which is when you must place other objects on the page or screen close to the table, such as additional tables, graphs, or text. When you can't surround the table with sufficient white space, a light boundary around the table may be useful to separate it from the objects around it, enabling the reader to focus on it without distraction from surrounding content.

Despite the uselessness of grids and the limited usefulness of rules to delineate columns and rows in tables, they are sometimes useful for grouping or highlighting subsets of data. Consider the following example, in which a column of totals has been included at the right, and additional white space has been used to group the totals as distinct from the other columns.

Product	Jan	Feb	Mar	Apr	May	Jun	Total
Product 01	93,993	84,773	88,833	95,838	93,874	83,994	541,305
Product 02	87,413	78,839	82,615	89,129	87,303	78,114	503,414
Product 03	90,036	81,204	85,093	91,803	89,922	80,458	518,516
Product 04	92,737	83,640	87,646	94,557	92,620	82,872	534,072
Product 05	83,733	75,520	79,137	85,377	83,627	74,826	482,220
Total	447,913	403,976	423,323	456,705	447,346	400,264	2,579,526

FIGURE 8.11 This example uses white space to make one column, in this case the column of totals at the right, stand out as different from the other columns.

Although extra white space to highlight the *Total* column is effective, if space is limited you could use a light rule to set this column apart instead:

Product	Jan	Feb	Mar	Apr	May	Jun	Total
Product 01	93,993	84,773	88,833	95,838	93,874	83,994	541,305
Product 02	87,413	78,839	82,615	89,129	87,303	78,114	503,414
Product 03	90,036	81,204	85,093	91,803	89,922	80,458	518,516
Product 04	92,737	83,640	87,646	94,557	92,620	82,872	534,072
Product 05	83,733	75,520	79,137	85,377	83,627	74,826	482,220
Total	447,913	403,976	423,323	456,705	447,346	400,264	2,579,526

FIGURE 8.12 This example uses a vertical rule to distinguish one column from the others, in this case the column of totals at the right.

Rather than the simple vertical rule above, we could have used four rules to form a border around the entire *Total* column. This would work fine, but it is a good practice to use the minimum means necessary to do the job. In the following example, the objective is to highlight the *March* and *Total* columns as particularly important, which cannot be done with vertical rules alone, so the combination of vertical and horizontal rules has been used to form borders around them:

Product	Jan	Feb	Mar	Apr	May	Jun	Total
Product 01	93,993	84,773	88,833	95,838	93,874	83,994	541,305
Product 02	87,413	78,839	82,615	89,129	87,303	78,114	503,414
Product 03	90,036	81,204	85,093	91,803	89,922	80,458	518,516
Product 04	92,737	83,640	87,646	94,557	92,620	82,872	534,072
Product 05	83,733	75,520	79,137	85,377	83,627	74,826	482,220
Total	447,913	403,976	423,323	456,705	447,346	400,264	2,579,526

FIGURE 8.13 This example uses vertical and horizontal rules to form borders around particular sets of data to highlight them.

When you use rules, be sure to subdue them visually in relation to the data by keeping the lines as thin and light as possible. I've found that shades of gray often work best and can be relied on to remain subtle in comparison to data rendered in black, even when photocopied in black and white.

FILL COLOR

When white space alone can't be used to effectively delineate columns and rows in tables, fill shades and hues work better than grids and rules. When subtly designed, fill colors are less distracting to the eye as it scans across them. They are limited, however, to one direction, delineating columns or rows but not both simultaneously. Here's an example of how the use of a light shade of gray on alternating rows can be used to aid scanning across long rows of data:

Product	Jan	Feb	Mar	Apr	May	Jun	Jul	Aug	Sep	Oct	Nov	Dec
Product 01	93,993	84,773	88,833	95,838	93,874	83,994	84,759	92,738	93,728	93,972	93,772	99,837
Product 02	87,413	78,839	82,615	89,129	87,303	78,114	78,826	86,246	87,167	87,394	87,208	92,848
Product 03	90,036	81,204	85,093	91,803	89,922	80,458	81,191	88,834	89,782	90,016	89,824	95,634
Product 04	92,737	83,640	87,646	94,557	92,620	82,872	83,626	91,499	92,476	92,716	92,519	98,503
Product 05	86,245	77,785	81,511	87,938	86,136	77,071	77,773	85,094	86,002	86,226	86,043	91,608
Product 06	88,833	80,119	83,956	90,576	88,720	79,383	80,106	87,647	88,582	88,813	88,624	94,356
Product 07	82,614	74,511	78,079	84,236	82,510	73,826	74,498	81,511	82,382	82,596	82,420	87,751
Product 08	85,093	76,746	80,421	86,763	84,985	76,041	76,733	83,957	84,853	85,074	84,893	90,384
Product 09	87,646	79,048	82,834	89,366	87,535	78,322	79,035	86,475	87,399	87,626	87,440	93,095
Product 10	90,275	81,420	85,319	92,047	90,161	80,672	81,406	89,070	90,021	90,255	90,063	95,888

FIGURE 8.14 This example uses a fill shade of light gray to assist horizontal scanning.

As you can see, it doesn't take much. This shade of gray is very light yet visible enough to assist our eyes in scanning across a single row without confusing it with the rows above or below. The shading is light enough that vertical scanning down a column is not significantly disrupted. Here's another example of the same practice, this time used to delineate columns:

Product	Jan	Feb	Mar	Apr	May	Jun	Jul	Aug	Sep	Oct	Nov	Dec
Product 01	93,993	84,773	88,833	95,838	93,874	83,994	84,759	92,738	93,728	93,972	93,772	99,837
Product 02	557,378	502,704	526,780	568,319	556,673	498,084	502,621	549,936	555,807	557,254	556,068	592,033
Product 03	574,100	517,785	542,583	585,369	573,373	513,027	517,699	566,434	572,481	573,972	572,750	609,794
Product 04	304,273	274,426	287,569	310,246	303,888	271,904	274,381	300,210	303,415	304,205	303,557	323,191
Product 05	30,427	27,443	28,757	31,025	30,389	27,190	27,438	30,021	30,342	30,420	30,356	32,319
Product 06	305,186	275,249	288,432	311,176	304,799	272,720	275,204	301,111	304,325	305,118	304,468	324,161
Product 07	6,104	5,505	5,769	6,224	6,096	5,454	5,504	6,022	6,087	6,102	6,089	6,483
Product 08	61,220	55,215	57,859	62,422	61,143	54,708	55,206	60,403	61,048	61,207	61,076	65,027
Product 09	50,813	45,828	48,023	51,810	50,748	45,407	45,821	50,134	50,670	50,801	50,693	53,972
Product 10	52,337	47,203	49,464	53,365	52,271	46,770	47,196	51,638	52,190	52,326	52,214	55,591

FIGURE 8.15 This example uses a fill color of light gray to assist in vertical scanning.

Fill colors can also be used to group and highlight sections of data. In the example below, the intention is to simply delineate the headers and footers as distinct groups of information from the rest, not to highlight them as more important:

Product	Jan	Feb	Mar	Apr	May	Jun	Total
Product 01	93,993	84,773	88,833	95,838	93,874	83,994	541,305
Product 02	87,413	78,839	82,615	89,129	87,303	78,114	503,414
Product 03	90,036	81,204	85,093	91,803	89,922	80,458	518,516
Product 04	92,737	83,640	87,646	94,557	92,620	82,872	534,072
Product 05	83,733	75,520	79,137	85,377	83,627	74,826	482,220
Total	447,913	403,976	423,323	456,705	447,346	400,264	2,579,526

FIGURE 8.16 This example uses a fill color to group particular data.

Because we don't want to suggest that the names of the months or the totals are more important than the other data, I've used a subtle fill color.

In the example below, fill color has been used to highlight two cells of data:

Product	Jan	Feb	Mar	Apr	May	Jun	Total
Product 01	93,993	84,773	88,833	95,838	93,874	83,994	541,305
Product 02	87,413	78,839	82,615	89,129	87,303	78,114	503,414
Product 03	90,036	81,204	85,093	91,803	89,922	80,458	518,516
Product 04	92,737	83,640	87,646	94,557	92,620	82,872	534,072
Product 05	83,733	75,520	79,137	85,377	83,627	74,826	482,220
Total	447,913	403,976	423,323	456,705	447,346	400,264	2,579,526

FIGURE 8.17 This example uses a fill color to highlight particular data.

When your purpose is to highlight, the fill color should be more noticeable. Don't go overboard, though, unless you want your readers to pay attention to nothing else.

Arranging Data

Our focus in this section is on how we can best arrange the data in a table to tell the story. Should the categorical subdivisions be displayed vertically in a column or horizontally across a row? Should the data be separated into sub-groups with breaks in between? Should the data be sorted in a particular order?

COLUMNS OR ROWS

Categorical subdivisions can be freely arranged across columns and down rows. Some arrangements, however, work much better than others. When you begin to construct a table, you have already identified one or more categories that are required to communicate your message. You must now arrange the sets of categorical subdivisions, either across the columns or down the rows, to best tell the story. How do you determine what goes where? What questions should you ask yourself about the data to make that determination? Take a minute right now to put yourself in this position, using a real or hypothetical table. Focus on what you need to know about each set of categorical subdivisions to decide whether it should be arranged across the columns or down the rows.

. . . .

Here are the questions that I ask myself about each set of categorical data:

- How many subdivisions does it contain?
- What is the maximum number of characters in any one of its subdivisions?
- Do the subdivisions involve a time-series or ranking relationship?

If the answer to the question "How many subdivisions does it contain?" is "only a few," then you have the option of arranging them across the columns. If there are more than a few, you have no choice but to arrange them in a single column because of the limited width of a standard page or screen. For instance, because there are only four sales regions in the following table, they can be arranged across the columns, with a separate column for each region:

Product	Regions			
	North	East	South	West
Product 01	94	152	174	87
Product 02	122	198	226	113
Product 03	101	164	188	94
Product 04	142	230	263	131
Product 05	132	214	244	122
Product 06	174	282	323	161
Product 07	401	648	742	371
Product 08	281	454	519	260
Product 09	112	182	208	104
Product 10	584	944	1,081	540
Product 11	543	878	1,005	502
Product 12	163	263	301	151
Product 13	489	790	904	452
Product 14	327	529	606	303
Product 15	295	476	545	273
Total	3,960	6,403	7,330	3,665

FIGURE 8.18 This is an example of arranging categorical subdivisions, in this case sales regions, as separate columns.

The same may not true of the products in this example, however, because it would be more difficult to fit the 15 columns that they would require across the width of the page or screen. If you had 15 sales regions as well, you would need to display both the product and region data down the rows, perhaps as follows:

Region	Product	Units Sold
Region 01	Product 01	152
	Product 02	198
	Product 03	164
	Product 04	230
	Product 05	214
	Product 06	282
	Product 07	648
	Product 08	454
	Product 09	182
	Product 10	944
	Product 11	878
	Product 12	263
	Product 13	790
	Product 14	529
	Product 15	476
Region 02	Product 01	443
	Product 02	133
	Product 03	399

FIGURE 8.19 This is an example of arranging all categorical subdivisions in single columns down the rows, because each set has too many subdivisions to arrange across the columns.

If you compare the last two examples, you may notice that by arranging both sets of categorical subdivisions side by side down the rows (*Figure 8.19*) rather than with one set down the rows and the other across the columns (*Figure 8.18*), you were forced to give up something that could be seen more readily in the bidirectional arrangement of the data. Can you see it? With the unidirectional arrangement of the data in *Figure 8.19*, you can easily compare the sales of different products within a given region, but it is more difficult to compare the sales of a product in different regions because those data aren't close together. With the bidirectional arrangement in *Figure 8.18*, the sales for the various regions are close to one another, making comparisons easier. This is worth noting. Bidirectional tables arrange more of the quantitative values closer together than unidirectional tables, offering a distinct advantage when comparisons need to be made. For this reason, you should almost always arrange the categorical subdivisions bidirectionally when you can fit one or more sets of categorical subdivisions across the columns.

The next question to consider is "What is the maximum number of characters in any one of the subdivisions?" If some of the subdivisions contain a large number of characters, this tells you that if you arrange them across the columns, the columns will have to be wide. Even if you have a small number of subdivisions, you may not have the horizontal space required to arrange them in wide columns. If you arrange the subdivisions down the rows in a single column, however, you only need a single wide column, which saves considerable horizontal space.

Next, you should ask the question, "Do the subdivisions involve a time-series or ranking relationship?" Why does it matter? Some sequential relationships are more easily understood when they are arranged in a particular way, either horizontally from left to right or vertically from top to bottom. If your table contains time-series data (years, quarters, months, etc.), what would usually work best, a left-to-right or top-to-bottom arrangement? We naturally think of time as moving from left to right, rather than down from top to bottom. Consequently, whenever space permits, time-series subdivisions should be arranged across the columns, as in the following example:

Region	1992 Q1	Q2	Q3	Q4	1993 Q1	Q2
North	393	473	539	639	439	538
East	326	393	447	530	364	447
South	401	483	550	652	448	549
West	538	647	737	874	601	736
Total	1,658	1,996	2,274	2,696	1,852	2,270

FIGURE 8.20 This example shows the preferred arrangement of time-series data across the columns from left to right.

A vertical arrangement would not work as well, as in the following:

Year	Qtr	Region North	East	South	West	Total
1992	1	393	326	401	538	1,658
	2	473	393	483	647	1,996
	3	539	447	550	737	2,273
	4	639	530	652	874	2,695
1993	1	439	364	448	601	1,852
	2	538	447	549	736	2,270

FIGURE 8.21 This example shows an awkward arrangement of time-series data.

When you display ranking relationships, such as your top 10 customers, the arrangement of subdivisions that works best in tables is vertical, from top to bottom, as in the following example:

Rank	Product	Sales (U.S. $)
1	Product J	1,939,993
2	Product E	1,784,794
3	Product G	1,642,010
4	Product A	1,510,649
5	Product D	1,389,797
6	Product C	1,278,614
7	Product B	1,176,324
8	Product H	1,082,219
9	Product F	995,641
10	Product I	915,990

FIGURE 8.22 This example shows a vertical arrangement of ranked data.

Even if you had room to arrange the 10 products horizontally across the columns, you generally wouldn't want to because rankings look more natural and communicate more effectively when arranged from top to bottom.

GROUPS AND BREAKS

It is often appropriate, and perhaps even necessary, to break sets of data into smaller groups. Sometimes you do so in order to direct your readers to examine groups of data in isolation from the rest. Sometimes you do so because smaller chunks of data at a time are easier to handle. Sometimes you are forced to break the data into groups simply because you can't fit everything horizontally across the page or screen. Sometimes you create group breaks so you can include summary values, such as sums and averages.

Whenever you break the data into smaller chunks and arrange them vertically, do so logically, based on the subdivisions of one or more categories. For instance, you may group categorical subdivisions of time expressed as years, starting a new group of data for each new year, or you may choose organizational divisions of the company, starting a new group of data for each division. You may need to break the data into even smaller chunks, basing the groups on a combination of categorical subdivisions, such as countries and sales regions within countries, as illustrated in the following example.

Country: USA   Region: North

Product	Jan	Feb	Mar	Apr	May	Jun	Jul	Aug	Sep	Oct	Nov	Dec
Product 01	93,993	84,773	88,833	95,838	93,874	83,994	84,759	92,738	93,728	93,972	93,772	99,837
Product 02	87,413	78,839	82,615	89,129	87,303	78,114	78,826	86,246	87,167	87,394	87,208	92,848
Product 03	90,036	81,204	85,093	91,803	89,922	80,458	81,191	88,834	89,782	90,016	89,824	95,634
Product 04	92,737	83,640	87,646	94,557	92,620	82,872	83,626	91,499	92,476	92,716	92,519	98,503
Product 05	86,245	77,785	81,511	87,938	86,136	77,071	77,773	85,094	86,002	86,226	86,043	91,608
Product 06	88,833	80,119	83,956	90,576	88,720	79,383	80,106	87,647	88,582	88,813	88,624	94,356
Product 07	82,614	74,511	78,079	84,236	82,510	73,826	74,498	81,511	82,382	82,596	82,420	87,751
Product 08	85,093	76,746	80,421	86,763	84,985	76,041	76,733	83,957	84,853	85,074	84,893	90,384
Product 09	87,646	79,048	82,834	89,366	87,535	78,322	79,035	86,475	87,399	87,626	87,440	93,095
Product 10	90,275	81,420	85,319	92,047	90,161	80,672	81,406	89,070	90,021	90,255	90,063	95,888
Total	$884,886	$798,085	$836,307	$902,255	$883,765	$790,751	$797,953	$873,070	$882,391	$884,688	$882,805	$939,903

Country: USA   Region: East

Product	Jan	Feb	Mar	Apr	May	Jun	Jul	Aug	Sep	Oct	Nov	Dec
Product 01	93,993	84,773	88,833	95,838	93,874	83,994	84,759	92,738	93,728	93,972	93,772	99,837
Product 02	87,413	78,839	82,615	89,129	87,303	78,114	78,826	86,246	87,167	87,394	87,208	92,848
Product 03	90,036	81,204	85,093	91,803	89,922	80,458	81,191	88,834	89,782	90,016	89,824	95,634

FIGURE 8.23 This example shows a grouping of data based on a combination of categories, in this case country and region.

Notice that if you want to sum sales by region, it would be awkward to incorporate them into the table without breaking the data into regional groups with group footers. Notice also that the group headers (e.g., *Country: USA Region: North*) interrupt vertical scanning of the information. This isn't a problem in this case, for the interruption reinforces the fact that a new group has begun and causes readers to briefly focus on the header information.

Whatever your reason for breaking the data into groups, keep in mind the following design practices:

- Use vertical white space between the groups but only enough to make the break noticeable. Excessive white space does nothing but spread the data farther apart and reduce the amount of information on the page or screen, which is rarely useful.
- Repeat the column headers at the beginning of each new group. This will help your readers keep track of the content in each column.
- Don't vary the structure of the table from group to group. For instance, each group should contain the same columns, in the same order, with the same widths; otherwise, your readers will be forced to relearn the structure at the beginning of each new group.
- When you group data based on multiple categories (e.g., country and region), position the group headers on the same row in order, from left to right, unless there isn't enough horizontal space to do so. This arrangement saves vertical space and clearly represents the hierarchical relationship between the categorical sets. *Figure 8.23,* on the previous page illustrates this arrangement.
- When you want your readers to examine each group of data in isolation from the rest, start each group on a new page.

These design practices are not arbitrary. Each of them is effective because it corresponds to the workings of visual perception.

COLUMN SEQUENCE

The primary considerations when you are determining the best order in which to arrange the columns of a table are the following:

- Each set of categorical subdivisions that is arranged down the rows of a single column should be placed to the left of quantitative values that are associated with it.
- If there is a hierarchical relationship between sets of categorical subdivisions (e.g., between product families and products), they should be sequenced from left to right to reflect that order.
- Quantitative values that are derived from another set of quantitative values using a calculation should generally be placed in a column just to the right of the column from which they were derived.

Let's look at an example of this last design practice before continuing the list:

Product	Units Sold	Actual Revenue	% of Total	Fcst Revenue	% of Fcst
Product A	938	187,600	47%	175,000	107%
Product B	1,093	114,765	28%	130,000	88%
Product C	3,882	62,112	15%	50,000	124%
Product D	873	36,666	9%	40,000	92%
Product E	72	2,088	1%	50,000	4%
Total	6,858	$403,231	100%	$445,000	91%

FIGURE 8.24 This example shows columns of quantitative values that are derived (i.e., calculated) from another column positioned to the right of the source column.

In this example, even though the *% of Total* column is somewhat ambiguous because it doesn't clarify what the percentages are based on (which it really should), most readers would still assume that it is the percentage of revenue for each product compared to the total revenue for all products. They would do so because this column is to the right of the *Actual Revenue* column. Even if its header were clearer (e.g., *% of Total Actual Revenue*), confusion would result if it were placed to the right of the *Units Sold* column or the *Fcst Revenue* column.

Now, back to our list of design practices.

- Columns containing sets of quantitative values that you want your readers to compare easily should be placed as close to one another as possible.

In the following example, last year's revenue has intentionally been placed next to this year's revenue to encourage comparisons between them:

Product	Units Sold	Last Year's Revenue	This Year's Revenue	Fcst Revenue	Planned Revenue
Product A	938	159,497	187,600	175,000	160,000
Product B	1,093	123,007	114,765	130,000	125,000
Product C	3,882	45,384	62,112	53,000	50,000
Product D	873	41,003	36,666	38,000	40,000
Product E	72	2,485	2,088	4,000	5,000
Total	6,858	$371,376	$403,231	$400,000	$380,000

FIGURE 8.25 This example places particular columns next to one another to encourage comparisons.

There is one more design practice that deserves a place on the list, but I want you to work a little for this one. Imagine that you are designing a table that will display units sold for a set of five products (i.e., products A through E) and a set of four sales channels (i.e., *Direct, Reseller, Distributor,* and *Original Equipment Manufacturer (OEM)*). You will have three columns: one for units sold, one for products, and one for sale channels. If you want to help your readers make easy comparisons between sales through the various channels for each product, in what order would you place the columns?

. . . .

To achieve this objective, you would place the sales channel column to the right of the product column. Here's an example of how it might look, with an additional column of percentages thrown in to enhance the comparisons:

Product	Channel	Units Sold	% of Total
Product A	Direct	8,384	26.53%
	Reseller	7,384	23.37%
	Distributor	10,838	34.30%
	OEM	4,993	15.80%
Product B	Direct	5,939	23.46%
	Reseller	7,366	29.10%
	Distributor	8,364	33.04%
	OEM	3,645	14.40%

FIGURE 8.26 This example arranges columns containing categorical subdivisions in a particular sequence to support comparisons that are important to the message.

Here's a statement of this design practice:

- To enable easy comparisons between individual members of a particular set of categorical subdivisions, either arrange them across multiple columns if space allows or in a single column to the right of the other columns of categorical subdivisions.

That's quite a mouthful, but I hope the example we worked through helps to make sense of it. In the example, because you wanted to help your readers make comparisons among sales through the various channels for individual products, the channel column had to go to the right of the product column. Placing the product column to the right of the channel column would not have served this purpose.

DATA SEQUENCE

The common term for sequencing data in a table is *sorting*. Numbers and dates both have a natural order that is meaningful. When you need to sequence numbers, their quantitative order, either ascending or descending, is the only useful way to sequence them. The same is true regarding the chronological order of dates and times. Other data, such as the names of things (e.g., products, customers, and countries) and other types of identifiers (e.g., purchase order numbers) have a conventional order based on alphabetical sequence, which is useful for look-up purposes but isn't meaningful otherwise. Alphabetical order is useful in tables for data that have no built-in meaningful order, such as the names of your customers, for this arrangement makes it easy for your readers to find any customers that they're looking for by scanning alphabetically through the list.

However, when a set of categorical subdivisions has a natural and meaningful order to the business, sorting them in alphabetical order would be unnatural. For instance, if your company does business in five countries—the United States, England, France, Australia, and Germany—and 80% of your business is done is the United States, and the percentage of business conducted in the other countries declines in the order in which they are listed above, to list them alphabetically in tables would make no sense. The point is simple: if subdivisions have a natural order that is meaningful to the business, sort them in that order. This will not only create a sequence that makes sense but will also place near one another subdivisions that your readers will often wish to compare.

Formatting Text

In this section we'll focus on aspects of text formatting that play a role in effective communication using tables. By text, I'm referring to all alphanumeric characters, including numbers and dates.

ORIENTATION

As speakers of English, we are accustomed to reading language from left to right, arranged horizontally. Any other orientation is more difficult to read, as a quick scan of the following should confirm:

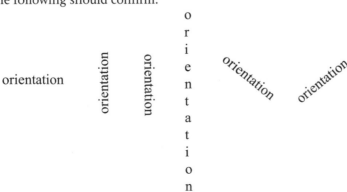

FIGURE 8.27 This figure illustrates various ways in which text can be oriented.

There is rarely a valid reason to sacrifice the legibility of text in a table by orienting it in any way other than horizontally from left to right. As we'll see in the next chapter, an alternate orientation is sometimes required in graphs, but this is seldom the case in tables.

ALIGNMENT

The most effective alignment of text is primarily a matter of convention rather than the mechanics of visual perception. The conventions at issue are powerful, so ignoring them leads to inefficiency, frustration, and confusion. The best practices of alignment can be summarized as follows:

- Numbers that represent quantitative values, as opposed to those that are merely identifiers (e.g., customer numbers), should always be aligned to the right.
- Dates are best aligned to the left, using a format that keeps the number of characters in each portion of the date (i.e., month, day, and year) constant.
- Text that expresses neither numbers nor dates works best when aligned to the left.

Because of the way numbers are written and read, aligning them to the left or center makes them difficult to interpret, as illustrated in the following example:

Sales	Sales	Sales
93,883.39	93,883.39	93,883.39
5,693,762.32	5,693,762.32	5,693,762.32
483.84	483.84	483.84
674,663.39	674,663.39	674,663.39
548.93	548.93	548.93
3,847.33	3,847.33	3,847.33
$6,467,189.20	$6,467,189.20	$6,467,189.20

FIGURE 8.28 These examples show the difficulty created when numbers are not aligned to the right.

This preference holds true for all units of measure, including currencies and percentages. You may be concerned that this preference for aligning numbers to the right would not hold true if the values in a particular column contained decimal digits of varying number, in which case the decimal points would not be aligned. It is best in such cases is to align both the decimal point and the final digit to the right. This can be accomplished by expressing each value using the same number of decimal digits, even when they are zeroes, as in the following example:

Rate
3.500%
12.675%
5.000%
13.250%
2.750%
13.125%
8.383%

FIGURE 8:29 This example shows the right alignment of numbers both to the decimal point and the last digit.

There is no strict convention for the alignment of dates. People differ in their preferences for left, right, and center alignment. From the perspective of reading efficiency, preference varies depending on the format of the date. I prefer left alignment accompanied by a format that maintains a constant number of digits in the date (e.g., 01/02/03 rather than 1/2/03). With this method, though the dates are left aligned in the column, they are in fact aligned along their right edges as well. The following example illustrates some of the alternatives, with my preference highlighted in gray:

Date	Date	Date	Date
12/17/02	12/17/02	12/17/02	12/17/02
1/2/03	01/02/03	1/2/03	01/02/03
1/17/03	01/17/03	1/17/03	01/17/03
2/9/03	02/09/03	2/9/03	02/09/03
10/29/03	10/29/03	10/29/03	10/29/03
12/1/03	12/01/03	12/1/03	12/01/03
1/1/03	01/01/03	1/1/03	01/01/03

FIGURE 8.30 These examples show the left and right alignment of dates, using two different date formats.

When we use a consistent number of digits in each date, each part of the date (month, day, and year) is aligned as well, making comparisons of a specific component (e.g., year) easy and efficient.

Text that expresses neither numbers nor dates works best when aligned to the left because of the historical conventions of printing. This includes numbers that are used as identifiers, rather than quantitative values, such as customer numbers. Here are examples of various text alignments:

Product Code	Product Name	Product Code	Product Name	Product Code	Product Name
A1838	2-Door Sport	A1838	2-Door Sport	A1838	2-Door Sport
A89	4-Door Sport	A89	4-Door Sport	A89	4-Door Sport
J98488	2-Door Luxury	J98488	2-Door Luxury	J98488	2-Door Luxury
J3883	4-Door Luxury	J3883	4-Door Luxury	J3883	4-Door Luxury
K9288	2-Door Truck	K9288	2-Door Truck	K9288	2-Door Truck
K38733	4-Door Truck	K38733	4-Door Truck	K38733	4-Door Truck

FIGURE 8.31 These examples show various text alignments, including the preferable left alignment highlighted in gray.

Don't be tempted by the aesthetic appeal of centered text in the columns of tables. The ragged left edge of the text makes scanning less efficient than the consistent leading edge of left-aligned text. I do, however, find that one exception to the practice of left alignment works well for columns of text: when the text entries each consist of the same number of characters and the column header consists of several more characters than the text entries. Here are two examples using columns with single character text entries, which display left and center alignment:

Cust Code	Preferred?	Cust Code	Preferred?
193847394	Y	193847394	Y
109388484	N	109388484	N
187466463	N	187466463	N
174563553	N	174563553	N
175357736	Y	175357736	Y
167374565	Y	167374565	Y

FIGURE 8.32 Theses examples show left- versus center-aligned text for columns with lengthy headers and single-character text entries.

Centering rather than left aligning such text entries does not produce a ragged left edge, so reading is not impaired. Centering also has the advantage of automatically adding more white space to the left and right of the text entries, thus distinguishing them more clearly from data in the adjacent columns. Dates may also be center aligned when their widths are formatted so they don't vary from date to date and the column header is significantly wider than the dates.

You may have noticed that in each of the examples above, the headers are aligned with the associated data. If the column's data are left aligned, its header is left aligned as well, and so on. This is intentional. This practice clearly establishes at the top of each column the nature of its alignment, thereby informing readers how to scan the data. The only exception that is sometimes useful involves *spanner headers*. When a header is used to label multiple columns, centering it across those columns often helps to clarify the fact that it refers to all of those columns rather to just a single column. This is illustrated by the centering of the header *Regions* in the following example:

Product	North	East	South	West	Total
Product A	94	152	174	87	507
Product B	122	198	226	113	659
Product C	101	164	188	94	547
Product D	142	230	263	131	766
Product E	132	214	244	122	712
Product F	304	491	562	281	1,638
Total	895	1,448	1,657	829	4,829

Regions spans North, East, South, and West.

FIGURE 8.33 This example shows a spanner header as an exception to the normal rules of alignment.

NUMBER AND DATE FORMAT

The display format that works best for numbers and dates is the one that exhibits the following characteristics:

- Includes no unnecessary information (e.g., excessive levels of precision, which we'll examine in the next section)
- Expresses the data using the format that is most familiar to your readers
- Most consistently aligns the data from row to row for efficient scanning

Rather than list and comment on all of the variations of number and date formats that exist, it will be more useful to highlight only those formatting practices that produce effective communication.

Here's a list of the useful practices for formatting numbers in tables:

- Place a comma to the left of every three whole-number digits (e.g., *1,393,033* rather than *1393033*).
- Truncate the display of whole numbers by sets of three digits to the nearest thousand, million, billion, etc., whenever numeric precision can be reduced without the loss of meaningful information, and declare that you've done so in the title or header (e.g., *U.S. dollars in thousands*).
- Use either a negative sign or parentheses to display negative numbers (e.g., *-8,395.37* or *(8,395.37)*), but if you use parentheses, keep the numbers that are enclosed in them right aligned with the positive numbers.
- Place a percentage sign (i.e., %) immediately to the right of every percentage value (e.g., *8.75%*).

The first three practices in this list are each firmly rooted in what you now know about visual perception. The commas break the numbers up into smaller chunks that can be stored more easily in short-term memory. Truncating whole numbers to remove everything below 1,000, everything below 1,000,000, etc., reduces the amount of information that must be read, resulting in greater reading efficiency, and reduces the amount of horizontal space required by the column. Aligning the right-most digits of positive and negative numbers, even when the negative numbers are enclosed in parentheses, keeps the numbers aligned for seamless scanning down the column. Look at the difference between the two formats below:

```
  83,743        83,743
2,339,844     2,339,844
 (67,909)      (67,909)
  60,036        60,036
 376,003       376,003
3,974,773     3,974,773
(576,533)     (576,533)
 937,764       937,764
     343           343
```

FIGURE 8.34 This example shows two alignment techniques for negative numbers that are enclosed within parentheses.

Your readers' eyes would be forced to jump back and forth slightly from row to row when scanning the values in the right-hand column in the example above.

Percentage signs should appear immediately to the right of every percentage value because percentages are used less often than other units of measure, so it's easy when reading down long columns of numbers to forget that you're looking at percentages. Including the percentage sign with each value takes care of this. Because currency values (e.g., U.S. dollars) are more common in business reports than percentages, there is no similar need to include the currency sign with each value. The conventional practice of reserving the currency sign for values appearing in summary rows (e.g., rows that display totals) works fine.

Now, here's a list of useful practices for formatting dates in tables:

* Express months either as a two-digit number (e.g., *02* rather than *2* for February) or a three-character word (e.g., *Feb* rather than *February*).
* Express days using two digits (e.g., *01* rather than *1* for the first day of the month).
* Use a format that excludes portions of the date that provide more precision than necessary.

The first two practices are related to the advantages of alignment for efficient visual perception. The last is related entirely to the level of precision that is appropriate, which we'll cover next.

If your readers are distributed internationally, keep in mind that conventional date formats differ in various parts of the world. The difference that is most prone to create confusion is that in the United States we generally list the month first, then the day, but it is common in Europe to list the day first, then the month. This isn't confusing when you express months as words rather than numbers (e.g., *Dec* rather than *12*), but when months are expressed as numbers using a format like *01-12-03*, the positions of the month and day are ambiguous. When this is the case, be sure to clarify the positions of the month and day portions of the date in your header, using a method like the following:

```
Order Date
(mm/dd/yy)
12/17/02
01/02/03
01/17/03
```

FIGURE 8.35 This example includes information in the column header to clarify the date format.

NUMBER AND DATE PRECISION

The *precision* of a number or a date is the degree of detail that it communicates. The number *12.825* represents a quantitative precision of three decimal digits while the number *12* only displays precision to the nearest whole number. The date *December 15, 2003* represents precision to the day level while the date *2003* only displays year-level precision. Selecting the appropriate level of precision for numbers and dates in tables boils down to a single design practice:

> The level of precision should not exceed the level needed to serve your communication objectives and the needs of your readers.

If the purpose of the table is to communicate the reconciliation of financial accounts, you had better display precision down to the penny (i.e., two decimal digits). If, however, you are presenting a multi-year comparison of sales revenue to the executives of a multi-billion dollar corporation, precision to the nearest million dollars may be appropriate. Forcing them to read six more digits of precision (or eight more digits if you include cents) would waste their time. If you express numbers with less precision than is available, you should always be careful to clearly state this in the title or header (e.g., *Rounded to the nearest* ꟷꟷꟷꟷ ꟷꟷꟷꟷ)

We are faced with choices regarding numeric precision whenever we produce new numbers as a result of calculations. When you divide 100 by 49 using spreadsheet software, you will likely get a result like 2.040816327. It's obvious that, for normal business purposes, precision to the level of nine decimals digits is excessive. Here are examples of the excessive levels of precision you would typically get using spreadsheet software to divide several common numbers by 100:

```
100 /  15 = 6.666666667
100 /  20 = 5.000000000
100 /  42 = 2.380952381
100 /  49 = 2.040816327
100 /  50 = 2.000000000
100 /  55 = 1.818181818
100 /  60 = 1.666666667
100 / 175 = 0.571428571
```

FIGURE 8.36 This example shows a level of precision that exceeds what is needed for most business purposes.

It may seem obvious that you would rarely need nine digits of decimal precision, but how many digits do you need? The answer depends on the message you're trying to communicate. For many purposes, rounding to the nearest whole number works fine. Here are three sample levels of precision for the list of numbers in the example above:

0 Decimals	1 Decimal	2 Decimals
7	6.7	6.67
5	5.0	5.00
2	2.4	2.38
2	2.0	2.04
2	2.0	2.00
2	1.8	1.82
2	1.7	1.67
1	0.6	0.57

FIGURE 8.37 These examples show three different levels of precision for the same set of values.

It isn't likely that more than a single decimal digit of precision would be significant in this case. In the example below, revenue is displayed as whole dollars per region along with each region's percentage contribution to the whole. The percentages have been repeated in four separate columns to illustrate different levels of precision:

Region	Revenue	% of Total	% of Total	% of Total	% of Total
Americas	636,663,663	40%	39.8%	39.82%	39.816%
Europe	443,874,773	28%	27.8%	27.76%	27.759%
Asia	399,393,993	25%	25.0%	24.98%	24.978%
Australia	99,838,333	6%	6.2%	6.24%	6.244%
Middle East	10,399,383	1%	0.7%	0.65%	0.650%
Africa	7,939,949	1%	0.5%	0.50%	0.497%
Total	$1,598,110,094	100%	100.0%	99.94%	99.944%

FIGURE 8.38 This example shows four different levels of precision for the display of the same set of percentage values.

If you needed in the example above to create a table that displays both a part-to-whole and a ranking relationship between the six regions using only percentages (i.e., excluding the actual revenue dollars), which level of precision should you select for the *% of Total* column?

．　．　．　．　．　．　．

Whole percentages (i.e., no decimal digits) would not differentiate the relative contributions or rank of the *Middle East* and *Africa* regions, and two decimal digits would provide more detail than necessary, so one decimal digit would probably work best.

Appropriate date precision is simpler to determine than numeric precision. Generally, you just need to decide whether your message requires precision to the level of year, quarter, month, or day. Once you determine which, select a date format that displays nothing below that level. Forcing your readers to read entire dates (e.g., *11/27/2003*) when all they need to know is the year (e.g., *2003*) not only slows them down, resulting in inefficiency and frustration, but also uses up more horizontal space, which may be in short supply.

FONT

There are hundreds of available fonts. A thorough knowledge of typography involves years of study and practice. Fortunately, we don't need to be typographic experts when designing tables. The best practices regarding font selection have two primary goals:

- The font should be as legible as possible.
- The same font should be used throughout.

Because our purpose in design is effective communication, fonts should be easy to read. Extravagant font choices intended to spice up the appearance of a table will reduce the efficiency of communication. This is seldom, if ever, an advantage in quantitative tables.

Fonts that are most legible tend to have a clean and simple design. This clean and simple appearance is not undermined by the use of *serifs*. A serif is a flourish, such as a line or curve, located at the end of a letterform. Fonts are generally classified into two primary types: 1) *serif* and 2) *sans-serif*. Many experts have argued that serif fonts are more legible, for the serif creates greater distinction between the individual characters. Whether this is true or not isn't entirely clear, even from a comprehensive review of the research. Serif and sans-serif fonts both work fine as long as you select one that is highly legible. The figure below shows examples of some of the most legible serif and sans-serif fonts as well as some that are not so legible:

Fine Legibility		Poor Legibility	
Serif	Sans-Serif	Serif	Sans-Serif
Times New Roman	Arial	*Script*	**Gill Sans Ultra**
Palatino	Verdana	**Broadway**	Papyrus
Courier	Tahoma	Old English	Tempus Sans ITC

FIGURE 8.39 These are examples of legible and poorly legible fonts, both of the serif and sans-serif types.

This list of legible fonts is not comprehensive by any means. It gives just a few examples of good ones that are readily available to most computer users.

Using the same font throughout a table is the best practice except when you must use the contrasting appearance of a different font either to group data or to make something stand out as different or more important. It is rare, though, that there isn't a better means to group or highlight particular content than the use of a contrasting font. For instance, the use of emphasis (e.g., boldfacing) is almost always preferable.

EMPHASIS AND COLOR

Text may be grouped or highlighted by using the attributes of emphasis and color. Emphasis, in this context, refers to boldfacing or italicizing the text. Bold-facing makes text heavier and darker, and italics make text stand out through a contrast in orientation, specifically by slanting the characters at an angle. Both color and emphasis can be used effectively to group and highlight data.

Summarizing Values

Tables are an ideal means to combine detail and summary values, simultaneously displaying both the big and the small pictures. A single value can be used to aggregate an entire set of values that extend down a column or across a row. In Chapter 2, *Numbers Worth Knowing*, we examined several numbers that can be used in tables to aggregate data. Summary values give your readers an overview in a single glance, and if they wish, they can then dig into the detailed values. This is a very powerful path to understanding, moving from general to specific, macroscopic to microscopic, summary to detail. Summary measures that are most useful in tables include the following:

- Measures of *sum* (i.e., the simple addition of a set of values)
- Measures of *average* (primarily the *mean*; secondarily the *median*)
- Measures of *occurrence* (i.e., a *count* of the number of instances of a thing or event)
- Measures of *distribution* (primarily the *minimum* and *maximum* values to represent their *range* and occasionally their *standard deviation*)

COLUMN AND ROW SUMMARIES

Most summary measures that appear in tables are used to aggregate the values down an entire column or across an entire row. A value that aggregates an entire column of values is a *column summary*, which generally appears in a separate row beneath the last row of detailed values. A value that aggregates an entire row of values is a *row summary*, which generally appears in a separate column to the right of the last column of detailed values. The following example includes both:

Product	Jan	Feb	Mar	Apr	May	Jun	Total
Product 01	93,993	84,773	88,833	95,838	93,874	83,994	**$541,305**
Product 02	87,413	78,839	82,615	89,129	87,303	78,114	**$503,414**
Product 03	90,036	81,204	85,093	91,803	89,922	80,458	**$518,516**
Product 04	92,737	83,640	87,646	94,557	92,620	82,872	**$534,072**
Product 05	86,245	77,785	81,511	87,938	86,136	77,071	**$496,687**
Product 06	88,833	80,119	83,956	90,576	88,720	79,383	**$511,587**
Product 07	82,614	74,511	78,079	84,236	82,510	73,826	**$475,776**
Product 08	85,093	76,746	80,421	86,763	84,985	76,041	**$490,049**
Product 09	87,646	79,048	82,834	89,366	87,535	78,322	**$504,751**
Product 10	90,275	81,420	85,319	92,047	90,161	80,672	**$519,893**
Total	**$884,886**	**$798,085**	**$836,307**	**$902,255**	**$883,765**	**$790,751**	**$5,096,049**

FIGURE 8.40 This is an example of a table that includes both column and row summaries.

The column and row summaries in this example have been boldfaced simply to make them stand out in the example, not to suggest that you must highlight summary values.

Column and row summaries frequently convey useful information. Whenever you design a table, ask yourself whether column and/or row summaries would be meaningful and useful even if they weren't specifically requested. If the answer is "yes" and there is space, include them. Because they are located at the edges of the table, a reader can easily ignore them whenever they're not needed.

GROUP SUMMARIES

Group summary values are similar to column and row summary values, differing only in that they summarize meaningful subsets of values rather than the entire set of values throughout the column or row. When you group and break rows of data into subsets based on one or more categories (e.g., sales regions), it is easy to include summary values for each of those groups, as in the following example:

Region: North

Product	Jan	Feb	Mar	Apr	May	Jun
Product 01	93,993	84,773	88,833	95,838	93,874	83,994
Product 02	87,413	78,839	82,615	89,129	87,303	78,114
Product 03	90,036	81,204	85,093	91,803	89,922	80,458
Product 04	92,737	83,640	87,646	94,557	92,620	82,872
Product 05	86,245	77,785	81,511	87,938	86,136	77,071
Product 06	88,833	80,119	83,956	90,576	88,720	79,383
Product 07	82,614	74,511	78,079	84,236	82,510	73,826
Product 08	85,093	76,746	80,421	86,763	84,985	76,041
Product 09	87,646	79,048	82,834	89,366	87,535	78,322
Product 10	90,275	81,420	85,319	92,047	90,161	80,672
Total	**$884,886**	**$798,085**	**$836,307**	**$902,255**	**$883,765**	**$790,751**

Region: East

Product	Jan	Feb	Mar	Apr	May	Jun
Product 01	93,993	84,773	88,833	95,838	93,874	83,994
Product 02	87,413	78,839	82,615	89,129	87,303	78,114
Product 03	90,036	81,204	85,093	91,803	89,922	80,458

FIGURE 8.41 This is an example of group summaries, in this case per region.

These summaries are group versions of column summaries.

Summaries of individual values across multiple columns in a single row are placed in a column of their own, generally to the right of the values they summarize. Here's a typical example:

Product	Jan	Feb	Mar	Q1 Total	Apr	May	Jun	Q2 Total
Product 01	93,993	84,773	88,833	267,599	95,838	93,874	83,994	273,706
Product 02	87,413	78,839	82,615	248,867	89,129	87,303	78,114	254,547
Product 03	90,036	81,204	85,093	256,333	91,803	89,922	80,458	262,183
Product 04	92,737	83,640	87,646	264,023	94,557	92,620	82,872	270,048
Product 05	86,245	77,785	81,511	245,541	87,938	86,136	77,071	251,145
Product 06	88,833	80,119	83,956	252,908	90,576	88,720	79,383	258,679
Product 07	82,614	74,511	78,079	235,204	84,236	82,510	73,826	240,572
Product 08	85,093	76,746	80,421	242,260	86,763	84,985	76,041	247,789
Product 09	87,646	79,048	82,834	249,528	89,366	87,535	78,322	255,223
Product 10	90,275	81,420	85,319	257,014	92,047	90,161	80,672	262,879
Total	$884,886	$798,085	$836,307	$2,519,278	$902,255	$883,765	$790,751	$2,576,771

FIGURE 8.42 This is an example of group column summaries, in this case totaling monthly revenue by quarter.

These summaries are group versions of row summaries.

Extra care should be taken, when intermingling group summary columns with detail columns, to make a clear distinction between the two types. You certainly don't have to make them as distinct as I have in the above example (and should-n't unless you want them to stand out as more important than the detail columns) but you do need to make sure that your readers notice that the sum-maries and details are different. In the example below, where no visual distinc-tion is made between the detail and summary columns, readers might mistak-enly perceive the summaries as just another column of monthly values:

Product	Jan	Feb	Mar	Q1 Mo Avg	Apr	May	Jun	Q2 Mo Avg
Product 01	93,993	84,773	88,833	89,200	95,838	93,874	83,994	91,235
Product 02	87,413	78,839	82,615	82,956	89,129	87,303	78,114	84,849
Product 03	90,036	81,204	85,093	85,444	91,803	89,922	80,458	87,394
Product 04	92,737	83,640	87,646	88,008	94,557	92,620	82,872	90,016
Product 05	86,245	77,785	81,511	81,847	87,938	86,136	77,071	83,715
Product 06	88,833	80,119	83,956	84,303	90,576	88,720	79,383	86,226
Product 07	82,614	74,511	78,079	78,401	84,236	82,510	73,826	80,191
Product 08	85,093	76,746	80,421	80,753	86,763	84,985	76,041	82,596
Product 09	87,646	79,048	82,834	83,176	89,366	87,535	78,322	85,074
Product 10	90,275	81,420	85,319	85,671	92,047	90,161	80,672	87,626
Total	$884,886	$798,085	$836,307	$839,759	$902,255	$883,765	$790,751	$858,924

FIGURE 8.43 This is an example of group column summaries that are not visually distinct from the columns that contain detailed data.

This tendency to confuse the summary columns with the detail columns becomes more pronounced as the reader's eyes move farther down the table, away from the top where the only clue to the identity of the column is in the headers. You can create the minimum necessary visual distinction through the use of any one of the available visual attributes that we've already examined, including something as subtle as extra white space separating the summary columns from the detail columns. Be careful to keep the distinction subtle unless the summary columns deserve greater attention than the detail columns.

HEADERS VERSUS FOOTERS

Values that summarize rows don't necessarily have to appear below the rows that they summarize. Although it is common practice to place them in table or group footers, there is at times an advantage to placing them above the rows that they summarize. When might it be beneficial to place summary values before the information that they summarize rather than after it? Take a moment to imagine tables that you commonly use to see whether you can think of circumstances when the placement of summary values in headers would offer an advantage to your readers.

. . . .

When the summary values are more important to your message or to your readers than the details, and placing them below the details would make them harder and less efficient to find, it often makes sense to place them in headers even though this is less conventional. Here are two examples of how this can be arranged:

Region: North	$1,568,586	$1,414,719	$1,482,474	$1,599,376	$1,566,600	$1,401,719
Salesperson	Jan	Feb	Mar	Apr	May	Jun
Abrams, S	93,993	84,773	88,833	95,838	93,874	83,994
Benson, J	87,413	78,839	82,615	89,129	87,303	78,114
James, R	86,245	77,785	81,511	87,938	86,136	77,071
Wilson, O	86,704	78,199	81,944	88,406	86,594	77,481
Yao, J	89,305	80,545	84,403	91,058	89,192	79,805

FIGURE 8.44 This is an example of placing column summary values in a header, in this case in the same row as a group subdivision (i.e., the north region).

Region: North						
Total	$1,568,586	$1,414,719	$1,482,474	$1,599,376	$1,566,600	$1,401,719
Salesperson	Jan	Feb	Mar	Apr	May	Jun
Abrams, S	93,993	84,773	88,833	95,838	93,874	83,994
Benson, J	87,413	78,839	82,615	89,129	87,303	78,114
James, R	86,245	77,785	81,511	87,938	86,136	77,071
Wilson, O	86,704	78,199	81,944	88,406	86,594	77,481
Yao, J	89,305	80,545	84,403	91,058	89,192	79,805

FIGURE 8.45 This is an example of placing column summary values in a header, in this case in their own row beneath the group value, because there isn't enough room to place them on the same row.

Though these examples show group summary values, the same practice is just as effective with grand totals. You may earn your readers' gratitude by eliminating the requirement that they flip to the last page of the report to get the big picture. If you're concerned that they'll be thrown off by not finding summary values where they conventionally appear after the details, you may put the summaries there are well. Summary values are generally so useful that the little loss of space used to place them in both the headers and the footers is a small price to pay for the advantage of greater availability.

Giving Page Information

Because of short-term memory constraints, effective table design requires that certain information be repeated on each new page. Otherwise, your readers may lose track of information that they need to interpret the table as they move from page to page. Two types of information in tables should be repeated:

- Column headers
- Row headers

REPEATING COLUMN HEADERS

When tables extend onto multiple pages, the columns are no longer labeled after the first page unless you repeat the column headers on each. This wouldn't present a problem if short-term memory weren't so limited, but it is. The space required to repeat the column headers is insignificant compared to the benefit of improved communication.

REPEATING ROW HEADERS

I've found that in actual practice, even those of us who are careful to repeat column headers on each page seldom think to repeat the row headers as well, yet the problem is the same. In the example below, the data pertaining to each state (*Alabama, Arkansas*, etc.) is introduced by the state's name in the *State* column.

State	Order Date	Sales Volume	Sales Revenue
Alabama	05/01/03	432	215,568
	05/02/03	534	266,466
	05/03/03	466	232,534
	05/04/03	354	176,646
	05/05/03	456	227,544
	05/08/03	553	275,947
	05/09/03	465	232,025
	05/12/03	580	200,011
	05/15/03	501	249,999
	05/16/03	556	277,444
	05/17/03	623	310,877
	05/19/03	563	280,937
	05/22/03	675	336,825
	05/23/03	702	350,298
	05/24/03	658	328,342
	05/26/03	798	398,202
	05/29/03	801	399,699
	05/30/03	735	366,765
	05/31/03	802	400,198
Arkansas	05/01/03	201	100,299
	05/02/03	247	123,253
	05/03/03	245	122,255
	05/04/03	277	138,223
	05/05/03	203	101,297

FIGURE 8.46 This example shows a typical table that groups data, in this case by state.

Because all the rows pertaining to each state are grouped together, it isn't necessary to repeat the name of the state on each row. In fact, to do so would be inefficient, causing readers to scan, on each new row, information that they already know. Also, the infrequent presence of data in the state column alerts readers through an obvious visual cue when information for a new state has begun.

Now imagine that information for the state of Alabama continues for five pages. By the time that you reached the third page, you might have forgotten which state you were examining, especially if your attention were pulled away, even briefly. You would see something like the next example:

State	Order Date	Sales Volume	Sales Revenue
	05/04/03	354	176,646
	05/05/03	456	227,544
	05/06/03	556	277,444
	05/07/03	598	298,402
	05/08/03	553	275,947
	05/09/03	465	232,035
	05/10/03	434	216,566
	05/11/03	676	337,324
	05/12/03	589	293,911
	05/13/03	688	343,312
	05/14/03	701	349,799
	05/15/03	501	249,999
	05/16/03	556	277,444
	05/17/03	623	310,877
	05/18/03	456	227,544
	05/19/03	563	280,937
	05/20/03	367	183,133
	05/21/03	356	177,644

FIGURE 8.47 This is an example of a page that lacks useful row headers.

Get the picture? Repeating the current row header at the beginning of each new page costs you nothing but provides your readers a great deal of benefit.

Summary at a Glance

Topic	Practices
Delineating Columns and Rows	• Use white space alone whenever space allows. • When you can't use white space, use subtle fill colors. • When you can't use fill color, use subtle rules. • Avoid grids altogether.
Arranging Data	• Columns or rows • Arrange a set of categorical subdivisions across separate columns if they are few in number and the maximum number of characters in those subdivisions is not too large. • Arrange times-series subdivisions horizontally across separate columns. • Arrange ranked subdivisions vertically down the rows. • Groups and breaks • Use just enough vertical white space between groups to make breaks noticeable. • Repeat column headers at the beginning of each new group. • Keep table structure consistent from group to group. • When groups should be examined independently, start each on a new page. • Column sequence • Place sets of categorical subdivisions that are arranged down the rows of a single column to the left of the quantitative values associated with them. • Place sets of categorical subdivisions that have a hierarchical relationship from left to right to reflect that hierarchy.

Topic	Practices
Arranging Data *(continued)*	• Place quantitative values that were calculated from another set of quantitative values just to the right of the column from which they were derived. • Place columns containing data that should be compared close together. • Value sequence • Whenever categorical subdivisions have a meaningful order, sort them in that order.

Formatting Text	• Orientation
	  • Avoid text orientations other than horizontal, left to right.
	• Alignment
	  • Align numbers to the right, keeping the decimal points aligned as well.
	  • Align dates to the left using a format that maintains a constant width.
	  • Align all other text to the left.
	  • Center non-numeric data if they all have the same number of characters and the number of characters in the header is significantly greater.
	• Number formatting
	  • Place a comma to the left of every three whole-number digits.
	  • Truncate the display of whole numbers by sets of three digits whenever numeric precision can be reduced to the nearest thousand, million, billion, etc.
	  • When negative numbers are enclosed in parentheses, keep the negative numbers themselves right aligned with the positive numbers.
	  • Place a percentage sign immediately to the right of every percentage value.
	• Date formatting
	  • Express months either as a two-digit number or a three-character word.
	  • Express days as two digits.
	• Number and date precision
	  • Do not exceed the required level of precision.
	• Font
	  • Select a font that is legible, and use the same font throughout the table.
	• Emphasis and color
	  • Boldface, italicize, or change the color of fonts when useful to group or highlight.

Summarizing Values	• Make columns containing group summaries visually distinct from detail columns. • Place summaries in the group header if its rows extend down multiple pages.
Giving Page Information	• Repeat column headers at the top of each page. • Repeat current row headers at the top of each page.

PRACTICE IN TABLE DESIGN

Nothing helps learning take root like practice. You will strengthen your developing expertise in table design by working through a few real-world scenarios.

Exercise #1

The following table has been prepared for a regional sales manager for the purpose of tracking the quarter-to-date performance of her sales representatives, including their relative performance. Given this purpose, take a look at the table, and follow the instructions below.

Quarter-to-Date Sales Rep Performance Summary
Quarter 2, 2003 as of March 15, 2003

Sales Rep	Quota	Variance to Quota	% of Quota	Forecast	Actual Bookings
Albright, Gary	200,000	-16,062	92	205,000	183,938
Brown, Sheryll	150,000	84,983	157	260,000	234,983
Cartwright, Bonnie	100,000	-56,125	44	50,000	43,875
Caruthers, Michael	300,000	-25,125	92	324,000	274,875
Garibaldi, John	250,000	143,774	158	410,000	393,774
Girard, Jean	75,000	-48,117	36	50,000	26,883
Jone, Suzanne	140,000	-5,204	96	149,000	134,796
Larson, Terri	350,000	238,388	168	600,000	588,388
LeShan, George	200,000	-75,126	62	132,000	124,874
Levensen, Bernard	175,000	-9,267	95	193,000	165,733
Mulligan, Robert	225,000	34,383	115	275,000	259,383
Tetracelli, Sheila	50,000	-1,263	97	50,000	48,737
Woytisek, Gillian	190,000	-3,648	98	210,000	186,352

List each of the problems that you detect in the design of this table:

Now, suggest a solution to each of these problems:

Exercise #2

The following table is used by mortgage brokers to look up the mortgage rates offered by several lenders. Brokers use this when they need to know the current rates offered by a particular lender for all of its loan programs. Given this purpose, take a look at the table and follow the instructions below.

Mortgage Loan Rates
Effective September 1, 2003

Loan Type	Term	Points	Lender	Rate
Adjustable	15	0	ABC Mortgage	6.0%
Adjustable	15	0	BCD Mortgage	6.0%
Adjustable	15	0	CDE Mortgage	6.0%
Fixed	15	0	ABC Mortgage	6.25%
Fixed	15	0	BCD Mortgage	6.75%
Fixed	15	0	CDE Mortgage	7.0%
Adjustable	30	.5	ABC Mortgage	6.125%
Adjustable	30	.5	BCD Mortgage	6.25%
Adjustable	30	.5	CDE Mortgage	6.5%
Fixed	30	.5	ABC Mortgage	6.5%
Fixed	30	.5	BCD Mortgage	7.0%
Fixed	30	.5	CDE Mortgage	7.25%
Adjustable	15	1	ABC Mortgage	5.675%
Adjustable	15	1	BCD Mortgage	5.675%
Adjustable	15	1	CDE Mortgage	5.75%
Fixed	30	1	ABC Mortgage	6.5%
Fixed	30	1	BCD Mortgage	6.5%
Fixed	30	1	CDE Mortgage	7.0%
Adjustable	15	1	ABC Mortgage	5.675%
Adjustable	15	1	BCD Mortgage	5.675%

List each of the problems that you detect in the design of this table:

Now, suggest a solution to each of these problems:

Exercise #3

The following table is used by the manager of the marketing department to examine the previous year's expenses in total and by quarter for each expense type. What you see below is the top portion of a page that is several pages into the table. The marketing manager finds this table frustrating. Can you help him out? Take a few minutes to respond to the instructions below:

2003 Marketing Department Expenses

Quarter	Transaction Date	Expense Type	Expense
	9/28/2003	Software	3837.05
	9/28/2003	Computer Hardware	10873.34
	9/29/2003	Travel	2939.95
	9/30/2003	Supplies	27.53
Qtr 4	10/1/2003	Supplies	17.37
	10/1/2003	Postage	23.83
	10/3/2003	Computer Hardware	3948.85
	10/3/2003	Software	535.98
	10/3/2003	Furniture	739.37
	10/3/2003	Travel	28.83
	10/4/2003	Entertainment	173.91
	10/15/2003	Travel	33.57
	10/16/2003	Membership Fees	395.95
	10/16/2003	Conference Registration	2195.00

List each of the problems that you detect in the design of this table:

Now, suggest a solution to each of these problems:

Exercise #4

It's now time to redirect your focus closer to home. Select three tables that are used at your place of business. Make sure that at least one of them is a table that you created. For each of the tables, respond to the following instructions:

Table #1

List each of the problems that you detect in the design of this table:

Now, suggest a solution to each of these problems:

Table #2

List each of the problems that you detect in the design of this table:

Now, suggest a solution to each of these problems:

Table #3

List each of the problems that you detect in the design of this table:

Now, suggest a solution to each of these problems:

Exercise #5

This remaining exercise asks you design a table from scratch to achieve a specific set of communication objectives. You may construct the table using any relevant software that is available to you, such as spreadsheet software. Imagine that you are a business analyst who was asked to assess the sales performance of your company's full line of 10 products during the preceding 12 months. During the course of your analysis, you discovered that the top two products account for more than 90% of total revenue and 95% of total profit. You believe it would be beneficial to either discontinue or sell off some of the worst-performing products. In order to make the case, you've decided to design a table that clearly communicates your findings. The table below provides the raw data that your findings were based on, which you can expand through calculations and arrange however you choose to construct your table. Dollars have been rounded to and expressed in thousands.

Product	Units Sold	Ext Cost (000s)	Ext Revenue (000s)
A	136	$3	$7
B	119	$59	$132
C	2,938	$7	$40
D	8	$54	$92
E	4,873	$387	$402
F	25,750	$760	$1,957
G	1,837	$395	$602
H	3	$15	$20
I	13,973	$3,298	$9,266
J	93	$2	$2

Once you've completed your table, take a few minutes to describe its design, including your rationale for each design feature, in the space below as needed.

You can find answers to the five exercises in Appendix F, *Answers to Practice in Table Design.*

9 GENERAL GRAPH DESIGN

The strong visual nature of graphs requires a number of unique design practices. The volume and complexity of quantitative information that you can communicate with a single graph are astounding but only if you recognize and avoid poor design practices that undermine your message.

> **Maintain visual correspondence to quantity**
> > **Correspondence to the tick marks**
> > **Zero-based scales**
> **Avoid 3-D**
> > **Data objects with 3-D depth**
> > **Graphs with 3-D depth**

The visual nature of graphs allows them to tap into the incredible power of visual perception to communicate quantitative information. When the message of a graph is contained in the patterns formed by its quantitative relationships, the graph will communicate elegantly but only if we avoid the far-too-common pitfalls of ineffective design.

We've already covered the aspects of quantitative communication that apply to both tables and graphs. None is more important to the design of graphs than the fundamental principle that was stated so eloquently by Edward Tufte: "Above all else show the data."[1] The quantitative message is encoded in the shape of the data, not in any characteristics of its container. The general practice of highlighting the data and subduing all else is even more important in the design of graphs than in the design of tables. Tables are a bit more forgiving of visual design flaws because tables encode data through the use of verbal language, visually displayed. Graphs, on the other hand, encode data as visual objects. These objects must be prominent, accurate, and clear.

Two fundamental principles of quantitative communication apply only to graphs:

- Encode quantities to correspond accurately to the visual scale.
- Avoid 3-D displays of quantitative data.

Both principles are firmly rooted in practical concerns; you can wreak havoc on communication if you ignore these principles.

Maintain Visual Correspondence to Quantity

You can only use two attributes of visual perception to reliably encode quantitative information: *line length* and *2-D position*. Quantitative values in graphs are either encoded visually as line length in the form of bars (i.e., the length of the

1. Edward R. Tufte (1983) *The Visual Display of Quantitative Information.* Cheshire CT: Graphics Press, page 92.

bar) or as 2-D position in the form of points and lines. Other visual attributes are either not perceived quantitatively at all (e.g., hue) or not well enough (e.g., 2-D area) to justify their use for quantitative encoding.

A line that is twice as long as another is perceived as having twice the quantitative value. Visual objects that encode quantitative values in graphs are interpreted by means of a scale line along the vertical or horizontal axis. When a bar that is twice as long as another corresponds to a value of 2 on the scale line, visual perception alone tells us that the value of the shorter bar should be around 1. If the shorter bar actually corresponds to a value of 1.75 or 0.5, something is amiss.

Sometimes graphs are intentionally manipulated to mask the truth contained in numbers. Darrell Huff, in his 1954 classic *How to Lie with Statistics*, was one of the earliest to express this concern. Advertisements are notorious sources of deliberately misleading graphs, but deception is not confined to advertising. You'll be faced many times with the temptation to manipulate graphs to give your case more strength than it deserves based on the actual numbers. Given the understanding of visual design that you are developing by reading this book, you will be even better equipped to manipulate visual design to exaggerate or hide the truth. It's easy to rationalize little design manipulations here and there to shade the truth slightly for a just cause. Be aware, though, that this manipulation does not qualify as design for communication. The goal of design for communication is always to promote an accurate understanding of the truth.

Here's a simple illustration of the potential for deliberate misinformation:

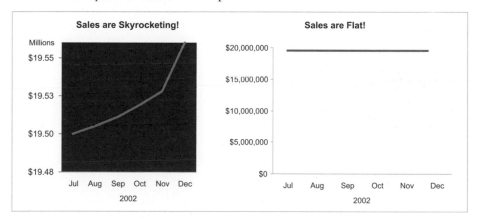

FIGURE 9.1 These two graphs display the same information in dramatically different ways, producing dramatically different messages.

The graph on the left has been deliberately manipulated to make an increase in sales from $19,500,000 in July to $19,560,000 in December, which is an increase of less than one-third of 1%, look like an increase of more than 200%. The graph on the right more accurately presents the data. Do you see the specific aspects of the graph on the left that were used to exaggerate the increase in sales? Take a moment to see how many you can find, and list them in the margin to the right.

• • • • • • •

Six design characteristics of the graph on the left are deliberate attempts to give the false appearance that sales have risen dramatically from July to December:

1. The scale on the Y axis does not start at zero. Rather, it starts at $19,475,000 and extends only to $19,560,000, thus making minor changes in sales appear extreme.

2. The plot area of the graph is taller than it is wide. This results in exaggerating changes along the Y axis, which provides the quantitative scale.

3. The line that encodes the quantitative values is green. The color green carries the meaning of growth and health in English-speaking cultures and dollars in the United States, so it reinforces the positive spin of the message. Also, placing the green line on a black background adds to its visual impact.

4. The bottom value label on the Y axis scale has been left off. This makes the fact that the scale doesn't begin with zero less obvious.

5. The highest value, which is the final value of $19,560,000, is set as the top of the scale. This gives the green line the appearance of extending right off the end of the graph in order to exaggerate its value.

6. Placement of the Y axis label *Millions* in the prominent upper left position near the title *Sales are Skyrocketing* is a deliberate attempt to suggest that they are increasing by millions.

This certainly represents a deliberate attempt to deceive, but I've seen worse. Can you think of any additional design changes that could be made to further hide the truth?

. . . .

Here's one that I've seen:

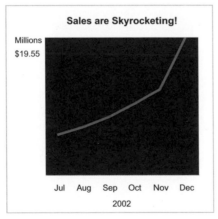

FIGURE 9.2 This is an extreme example of intentional deceit through graph design.

Notice the changes? All value labels except for one, along with all tick marks, have been removed from the Y axis scale. Without at least one more value on the scale, there is absolutely no way to know the extent of the increase in sales. The single starting value of $19,500,000, combined with the steep line of increasing value and the scale label *Millions*, work together to suggest a huge rate of increase.

Now, back to the principle that prompted our journey through the murky land of visual obfuscation. A quantity that is visually encoded in a graph should match the actual quantity that it represents. Two specific design practices will help you honor this correspondence:

- Make the distance between tick marks on a scale line correspond to the differences in the values that they represent.
- Generally include the value zero in your quantitative scale, and alert your readers when you don't.

Correspondence to the Tick Marks

You should always keep the distance between tick marks on a scale line consistent with the difference in the quantitative values that they represent. Software that generates graphs for you based on specified sets of values automatically enforces this practice. If the tick marks represent the values 1, 2, 3, 4, and 5, they will be positioned an equal distance from one another. If you ever produce graphs without the aid of graphing software, you should be sure to honor this practice. Approaching this from the opposite perspective, if you have a set of tick marks that are positioned at equal distances from one another, the values that you use to label them should also represent equal numeric intervals. Never place a gap in the values, such as in consecutive tick marks labeled as 1, 2, 7, 8, and 9, even if there are no values in the graph that fall within the missing range. To do so would undermine the graph's visual integrity.

You may recognize that these tick marks would not be equidistant if you were using something other than a standard scale, such as a logarithmic scale. We'll look at the special qualities and uses of logarithmic scales a little later.

Zero-Based Scales

When you set the bottom of your quantitative scale to a value greater than zero, differences in values will be exaggerated visually in the graph. You should generally avoid starting your graph with a value greater than zero, but when you need to provide a close look at small differences between large values, it is appropriate to do so. Make sure you alert your readers that the graph does not give an accurate visual representation of the values so that your readers can adjust their interpretation of the data accordingly. For instance, if the sales manager of the company with the subtly rising sales that we examined previously wanted to examine that rise in great detail, however insignificant it may be as a percentage increase, textual alerts similar to those found in the following example would be required:

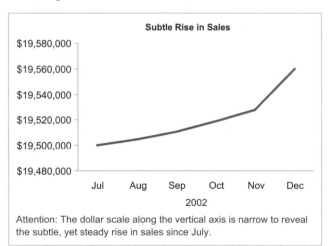

FIGURE 9.3 This is an example of an exception to the zero-base scale rule, which illustrates how such an exception must be clearly noted so that the distortion of the visually encoded values may be recognized and interpreted correctly.

Never eliminate zero from the quantitative scale, however, when bars are used to encode the values. Why? Because a bar encodes quantitative value primarily through its length, and, without zero as the base, the length will not correspond to its value.

When a graph represents both positive and negative numbers, zero will not mark the bottom of the scale, but it will still represent the base from which all values extend. The following two graphs contain the same set of positive and negative values. The graph on the left correctly displays zero as its base, but the one on the right mistakenly sets its lowest value as the base, resulting in a confusing and misleading representation of the values.

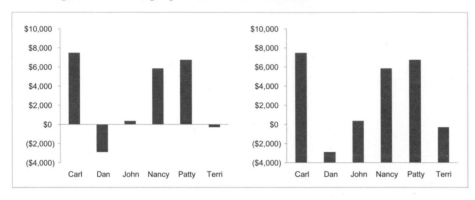

FIGURE 9.4 Both of these graphs display both positive and negative numbers. The graph on the left correctly sets zero as the base value, making it the point where the X axis intersects the Y axis. The graph on the right incorrectly sets the lowest value as the base.

Avoid 3D

When 3D is used in graphs, it takes one of two possible forms:

- Addition of the third dimension of depth to objects (e.g., bars) that are used to encode quantitative values, without the addition of a third scale line.
- Addition of the third dimension of depth to the overall graph, along with a third scale line.

Neither form is effective, but the reasons are entirely different.

Data Objects with 3-D Depth

We use three objects to encode quantitative values in graphs: points, bars, and lines. The addition of depth to a value-encoding object does not affect the object's value. Add depth to a series of bars, and what do you have? Nothing more than bars that now occupy more space and are harder to tie to values along the scale line. If you add depth to value-encoding points, like dots and squares, you get spheres and cubes that represent the same values as before, but now their depth makes it harder to align them accurately to the scale line. 3-D versions of lines look like ribbons, which suffer from the same problems.

Here are four variations of the same graph, three of which have added 3D to the bars:

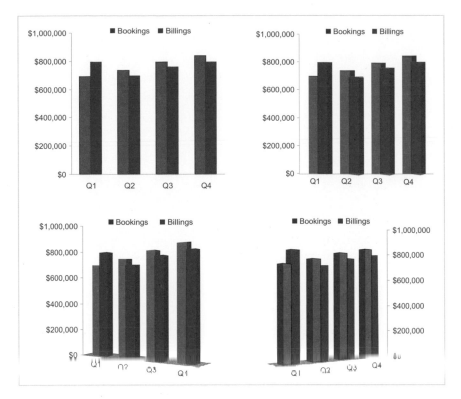

FIGURE 9.5 These four examples use the same values but display using bars in four different ways, three of which incorporate 3D.

Which graph do you find easier to read?

Most software makes it far too easy and tempting to add a third dimension to objects in graphs. This functionality is thrown in because people expect it, not because it's useful. It is far better to impress your readers with graphs they can easily understand and use, rather than graphs that look like video games and are difficult to interpret.

Remember the *data-ink ratio*. The addition of 3D to value-encoding objects adds ink, but not data. It adds meaningless visual content that your readers must take in and process, resulting in nothing but wasted time and effort.

Graphs with 3-D Depth

The third dimension of depth may be added to an entire graph through the use of a third axis. This third axis may be used for either a categorical or a quantitative scale. A categorical scale along the third axis allows you to add another set of categorical subdivisions that extend back along the axis, accompanied by related rows of quantitative values. A quantitative scale along the third axis allows you to either add depth to the existing bars, lines, or points, or another quantitative variable in a scatter plot. In theory, this is a valid way to include more information in a graph. In practice, it is simply too hard to read. Simulating 3-D space on a 2-D surface works nicely for paintings or technical illustrations, but quantitative values cannot be effectively communicated in this manner.

A few examples will vividly illustrate this point. Let's start with the same bookings information that we examined in the last set of examples, leaving out billings to keep matters simple.

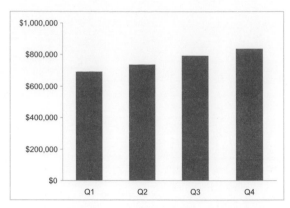

FIGURE 9.6 This is an example of a simple 2-D graph.

So far we have a very simple 2-D graph. Now let's say that we want to display these quarterly bookings by the four sales regions of north, east, south, and west. To do so, we could encode each region as a different hue and keep the graph 2D, as follows:

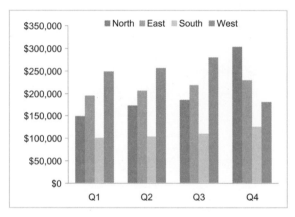

FIGURE 9.7 This 2-D graph has two sets of categorical subdivisions: quarters along the X axis and sales regions encoded as different hues.

This is still fairly easy to read. Rather than using hue to encode the four sales channels, however, we could add another axis to the graph, making it 3D, and display the sales channels along that axis.

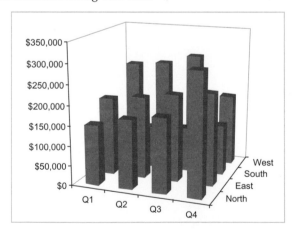

FIGURE 9.8 This is a simple 3-D graph, with sales in dollars along one axis, quarters along another, and sales regions along a third.

This is a very simple example of a 3-D graph with two categorical scales (quarters and regions) on one quantitative scale (dollars). What do you think? Does it work? Examine it for a moment, attempting to read and compare its values. Try to describe what makes this graph difficult to read.

. . . .

When a third dimension is added to a graph, adjustments are made by tilting, rotating, and adding perspective to the graph in order to make its data more visible. A 3-D display like this is called an *axonometric projection*. The previous example was tilted down 15 degrees, rotated clockwise 20 degrees, and given 30 degrees of perspective. These variables can be altered in an effort to make the graph easier to read. Even though the graph has been tilted and rotated in an attempt to make the rows of bars more visible, some bars will always remain partially or entirely hidden. Also, it's nearly impossible to line the bars up with values along the scale lines.

Software that generates 3-D graphs often includes walls and a floor to enhance the 3-D illusion, as well as grid lines on the walls in an effort to make the quantitative values easier to align with the scales lines. Here's the same graph as before with the addition of these features, along with black borders around the bars to more clearly delineate them:

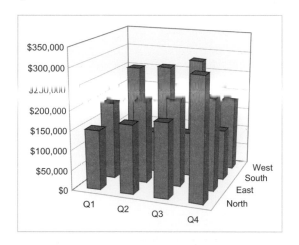

FIGURE 9.9 This is a 3-D graph that has been enhanced in an effort to make the values easier to read though the use of walls and a floor, grid lines on the walls, and borders around the bars.

Even though this is a fairly simple graph, these enhancements still don't solve the problems. Changing from the use of bars to lines to encode the data doesn't fix the problem either, as you can see in this example:

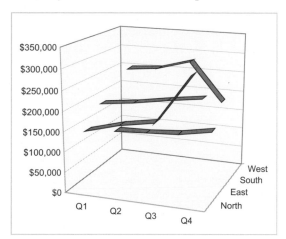

FIGURE 9.10 This graph displays the same data as *Figure 9.9* above but this time using lines to encode the values.

Support components called *drop lines* were invented to help us locate data objects in relation to scale lines, especially in 3-D graphs, but they clutter the graph and reduce its interpretation to a series of look-ups rather than perception of the data's shape. Here's the same graph, this time with drop lines:

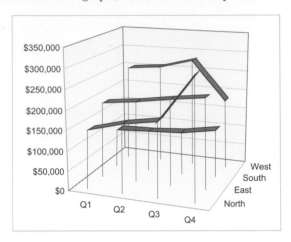

FIGURE 9.11 This graph is the same as the one in *Figure 9.10* with the addition of drop lines.

So far we've only examined the association of a categorical scale with the third axis. The problems don't get any better, however, when the third axis is used for a quantitative scale. Imagine a scatter plot that correlates employee salaries in dollars along one axis, tenure on the job in years along another axis, and level of education in years along the final axis. It's too difficult to tell where the points are positioned along the third axis.

3-D renderings of quantitative business information rarely work. Don't sacrifice effective communication through the use of 3-D fluff. Even when you are driven by a sincere desire to give your readers more information by using a third dimension, there are better ways to realize these good intentions. One effective technique is to use multiple related graphs in a series, which we'll explore in Chapter 11, *Design Solutions for Multiple Variables*.

Summary at a Glance

- Encode quantities to correspond accurately to the visual scale.
  - Keep the distance between tick marks on a scale line consistent with the difference in the quantitative values that they represent.
  - Generally include the value zero in your quantitative scale, and alert your readers when you don't. Always start the quantitative scale at zero when you use bars to encode the values.
- Avoid 3-D displays of quantitative data.

10 COMPONENT-LEVEL GRAPH DESIGN

A number of visual and textual components must work together in graphs to communicate quantitative information. If these components are out of balance or misused, the message suffers. For each component to serve its purpose effectively, you must understand its role and the design practices that enable it to fulfill its role with precision and grace.

> **Data component design**
> > **Points**
> > **Bars**
> > **Lines**
> > **Scale lines**
> > **Legends**
> > **Other text**
> **Support component design**
> > Axes and the data region
> > **Grid lines**

Like tables, graphs are constructed using components that fall into two categories:

- Data components
- Support components

Before we examine these components in detail, let's get our terminology straight. Here's a diagram that shows most of the terms used to describe graph components in this chapter:

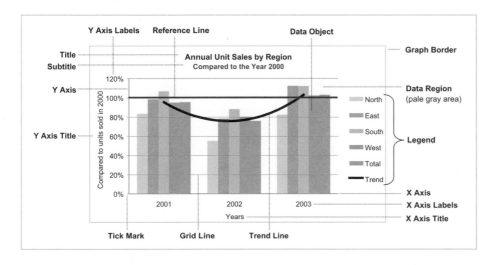

FIGURE 10.1 This diagram labels the components of a graph.

This figure is not meant to illustrate the best practices of graph design but simply to label the parts.

Data Component Design

Graphs communicate primarily through the channel of visual perception, tapping into our powerful ability to detect patterns in quantitative information, when visually displayed. Values are encoded in graphs as one or more of the following three types of objects:

- Points
- Bars
- Lines

To assign actual numbers to the values represented by these objects, graphs use quantitative scales along the axes, encoded as numeric labels and tick marks. Categorical subdivisions are primarily identified as visual attributes of the objects that encode the data values (e.g., color or location in 2-D space). Text labels are used to assign categorical subdivisions to these attributes, either along the axes or through the use of legends. Just like in tables, graphs may include additional text that can:

- Label
- Introduce
- Explain
- Reinforce
- Highlight
- Sequence
- Recommend
- Inquire

Because we've already spent some time in Chapter 5, *Fundamental Variations of Graphs*, examining how points, bars, and lines are commonly used to encode values, we'll focus now on the finer details of their design, then proceed to a handful of data components that we haven't examined so far.

Points

When you encode quantitative values as points (dots, squares, triangles, etc.), all the associated design practices address a single objective, which is to *keep each of the points distinctly visible*. Let's walk through some examples of graphs that illustrate the various visibility problems that can plague points.

Look at this first example and take a moment to describe its problem in the margin to the right:

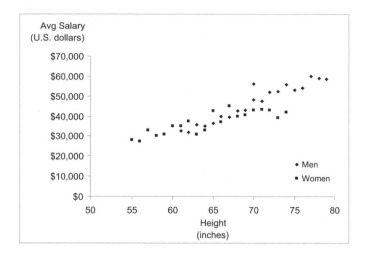

FIGURE 10.2 This is an example of points that suffer from a visibility problem.

The problem is that the points are so small that their distinct orientations (vertically oriented squares for the men versus squares rotated 45° (i.e., diamonds) for the women) cannot easily be distinguished. This lack of distinction between the two sets can be remedied in a number of ways. Take a minute and try to identify one or two ways that this problem may be fixed.

Here are the most effective remedies:

• Enlarge the points.

Nothing is different about the following example, except that the points have been enlarged:

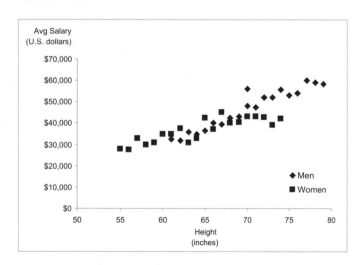

FIGURE 10.3 This example demonstrates that sets of points can be made more distinct by enlarging them.

Enlarging the points certainly helps, but here's another method that works even better:

- Select objects that are more visually distinct.

Here's an example:

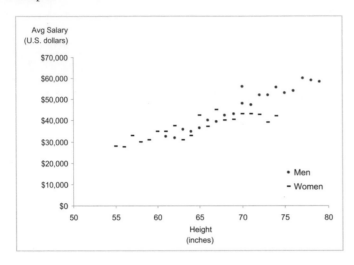

FIGURE 10.4 This example demonstrates that sets of points can be made more distinct from one another by using shapes that are very different.

In this example, the distinction has been clarified without enlarging the points at all but simply by selecting shapes that are more visually different from one another. Some shapes are more distinctive than others. Here's an example of one of the best attributes in this regard:

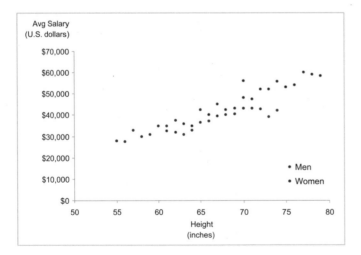

FIGURE 10.5 This example demonstrates that sets of points can be made more distinct from one another by using different hues.

The simple use of different hues makes it easy to see each set of points as distinct without any enlargement or change of shape. Different sizes, shapes, and hues are only three of the visual attributes that you can use to distinguish different sets of points. Of course, there's no rule that says you can't use more than one visual attribute at the same time to make the distinctions even more striking, as in the following:

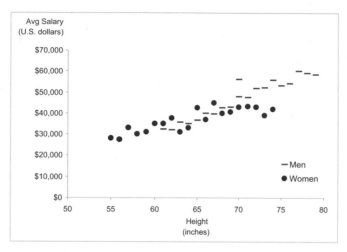

FIGURE 10.6 This example demon-strates that sets of points can be made very distinct from one another by varying multiple visual attributes simultaneously, in this case hue and shape, along with enlarging all the points.

Let's move on to the next visibility problem that can plague points, especially in scatter plots that show a large number of points. Take a look at the following example and describe in the right margin the problem that you detect:

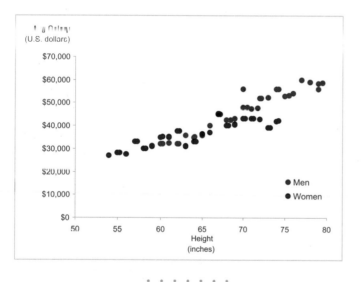

FIGURE 10.7 This example shows points that suffer from a visibility problem.

· · · · · · ·

This example has twice the number of points as the previous examples. The problem that it illustrates is that points can overlap each other, so that some of them may be obscured completely. When points are hidden or overlapped by other points to the degree that they cannot be distinguished, some of the information is invisible, resulting in inaccurate communication.

When points overlap somewhat, but none are entirely hidden, this can sometimes be remedied by enlarging the graph, decreasing the size of the points, or a combination of the two. These steps will reduce the overlap. If the problem persists, two more techniques can be used that usually fix the problem. Knowing what you now know about visual perception, look again at the last example and see whether you can come up with any methods to remedy the problems associated with the overlapping points.

· · · · · · ·

Any luck? Your ability to recognize opportunities to improve visual perception by manipulating visual attributes should be sharpening, so you may have detected that the solid nature of the points, the fact that they are filled with color, reduces their visibility when overlapping occurs. Let's see what happens when we remove the fill colors, leaving only the outlines of the dots:

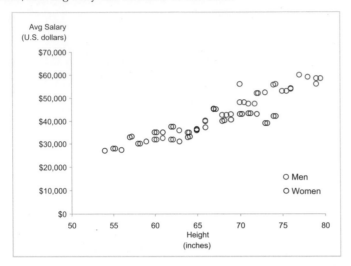

FIGURE 10.8 This example demonstrates that the problem of overlapping data points can be reduced by removing the fill color from the points.

This makes a big difference. When the points have transparent interiors, we can more readily see when points overlap. This same technique works with a variety of point shapes, not just dots.

When points belonging to different data sets obscure one another, one solution involves careful selection of the shapes of the points to make them more distinct from one another. Here's an example:

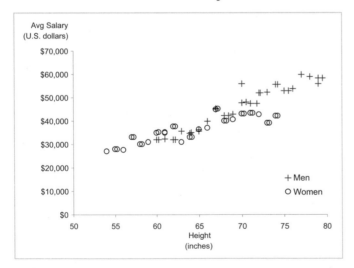

FIGURE 10.9 This example demonstrates that the problem of overlapping points can be reduced by selecting dramatically different shapes for the points.

Now, in addition to removing the problem caused by the fill color, we are intentionally selecting shapes that are dramatically distinct to represent the two sets of values. In fact, few simple shapes are as distinct from one another as circles (or ovals) and straight lines or crosses (i.e., plus signs or x's).

We have one more problem to identify and resolve related to the visibility of points. It is particular to the use of points and lines in combination to encode data. Take a look at the following example and identify the problem:

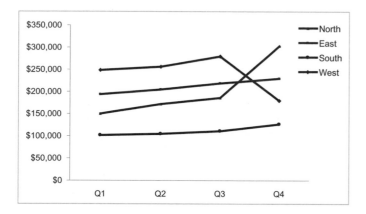

FIGURE 10.10 This is another example of a problem with the visibility of points, this time when they are connected with lines.

The problem here is fairly obvious. If you choose to use points and lines in combination, make sure that the points are not obscured by the lines. This is easy to remedy. When you combine points and lines, you usually want to emphasize the individual values marked by the points yet also want to display their connection and overall shape. If you wanted to emphasize the pattern of the values over the individual values, you wouldn't need the points at all. To emphasize the individual values while also providing a subtle sense of their shape, make the points visually prominent and the lines subdued in comparison. Here's an example:

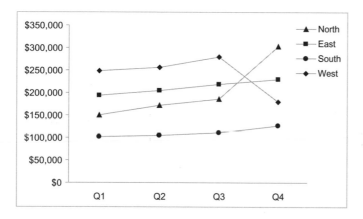

FIGURE 10.11 This example demonstrates a remedy to the problem of points that are obscured by the lines that connect them.

The design practices that we've uncovered regarding points may be summarized as follows:

- When sets of points cannot be clearly distinguished from one another, either enlarge them, select objects that are more visually distinct, or both.
- When points overlap, improve their visibility by enlarging the graph, reducing the size of the points, removing the fill colors, or some combination of these steps. If overlap between points belonging to different data sets is a problem, select radically distinct shapes, such as circles and crosses.
- When points and lines are used in combination, make the points visually prominent compared to the lines that connect them.

Bars

Several characteristics of bars deserve attention. In this section we'll examine the following:

- Orientation
- Proximity
- Fills
- Borders
- Bases

ORIENTATION

In this context, *orientation* refers to whether the bars run horizontally from left to right across the graph or vertically from bottom to top up the graph. Each orientation has advantages in particular circumstances.

Horizontal bars are the best choice when either of the following conditions exist:

- The graph displays a ranking relationship in descending order.
- The categorical labels on the bars won't fit side by side.

Because the purpose of a ranking relationship is to display a sequence of data, generally from high to low, a stacked arrangement of horizontal bars does the job especially well, as in the following example:

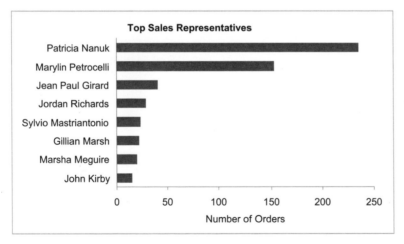

FIGURE 10.12 This graph displays a ranking relationship presented through the use of horizontal bars.

In this example, the top sales representative is represented by the top value in the graph, and so on, in order of rank. A left-to-right arrangement also works but with slightly less intuitive correspondence to ranking than top to bottom.

The example above introduces us to the second situation in which horizontal bars are preferable. Even though categorical labels can get much longer than these sales representatives' names, a great deal of horizontal space would be required to arrange them side by side in order to label vertical bars, as in the following example:

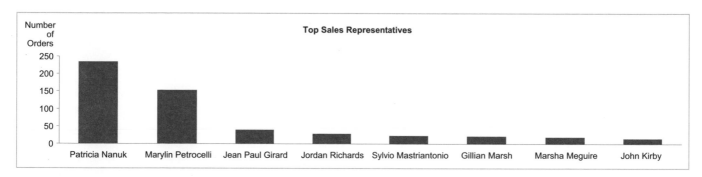

FIGURE 10.13 This graph illustrates the problem with vertical bars when their categorical labels are long.

Perhaps you could make this work, but what if the graph displayed 20 sales representatives rather than 10?

You may be thinking that this problem could be solved by orienting the names vertically or at least at an angle, as in the following:

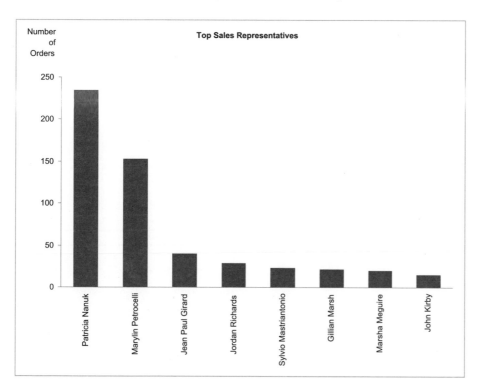

FIGURE 10.14 This graph illustrates the problem that results when we associate long categorical labels with vertical bars by orienting them vertically.

This solves the horizontal space problem but makes the names hard to read. When you must use vertical bars and the categorical labels are too long to fit side by side, opt for an upward-sloping angle of 45° or less rather than orienting the labels vertically. This is much easier to read, as in the following example:

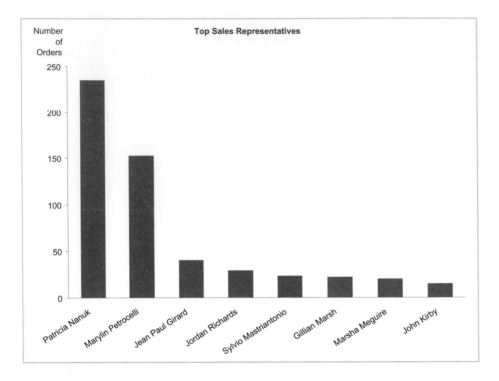

FIGURE 10.15 This graph orients the categorical labels at an angle, rather than vertically, to make them more legible. These are oriented at a 33° angle.

It is almost always better to solve the problem, however, with horizontal rather than vertical bars if you can.

PROXIMITY

Now let's consider how close bars should be placed to one another. You should always maintain a balance between the width of the bars themselves and the white space that separates them. In other words, rather than thinking of the space between the bars in absolute terms, think of it in terms of a ratio of the width of the bars to the width of the white space. Here are several examples of bars that are separated by different ratios of bar width to white space. Take a look at the next five figures and determine which seems to work best.

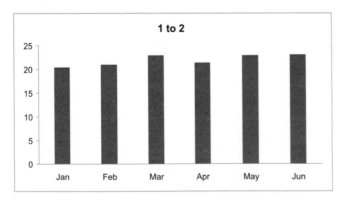

FIGURE 10.16 This is an example of a 1-to-2 ratio of bar width to intervening white space.

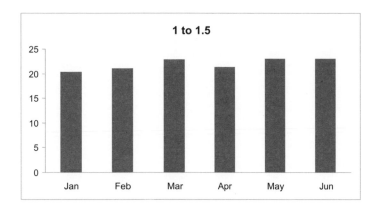

FIGURE 10.17 This is an example of a 1-to-1.5 ratio of bar width to intervening white space.

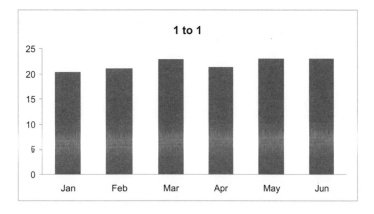

FIGURE 10.18 This is an example of a 1-to-1 ratio of bar width to intervening white space.

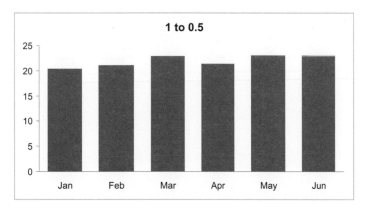

FIGURE 10.19 This is an example of a 1-to-0.5 ratio of bar width to intervening white space.

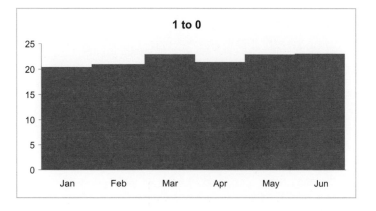

FIGURE 10.20 This is an example of a 1-to-0 ratio of bar width to intervening white space.

I am not aware of any research that suggests which of these ratios works best. Personally, I prefer to stick within the range extending from a ratio of 1:1.5 to 1:0.5 and lean toward a ratio of 1:1 as ideal. Larger ratios produce too much white space. At the other extreme, as in the 1:0 ratio example, the bars cease to be discrete, suggesting a continuous range of values that is only appropriate along an interval scale. At this extreme, the unique ability of bars to display individual values as discrete is almost entirely lost.

The primary situation when no space between bars works is when the bars that touch one another correspond to a different set of categorical subdivisions than the one labeled along the axis. Take a look at the two examples below:

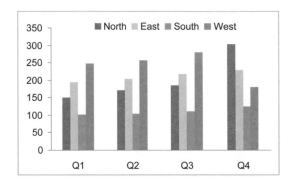

FIGURE 10.21 This graph illustrates when it is appropriate to place bars side by side without white space in between.

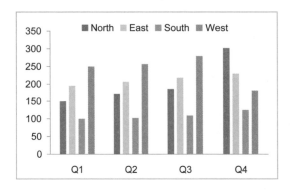

FIGURE 10.22 This graph demonstrates the inappropriateness of placing white space between bars when the bars correspond to different categorical subdivisions that are not encoded as position along an axis.

The first example lacks space between the contiguous bars, and the second exhibits a bar-to-white-space ratio of 1 to 0.5 When bars that are not labeled along the axis are grouped together (e.g., geographical regions in these examples), because they belong to the same categorical subdivision that is labeled on the axis (e.g., quarters), there is no need to insert white space between them. Because the bars for each region have already been distinctly rendered using different hues, further distinction through the use of white space isn't needed.

What about overlapping the bars? You could go further than removing white space between bars and actually overlap them. Examine the next example to see how this looks:

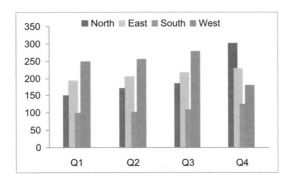

FIGURE 10.23 This is an example of overlapping bars, which illustrates the visual problems that overlapping creates.

When bars are overlapped in this manner, the result looks a bit like a jigsaw puzzle. The dark tan bars above appear more dominant than the others because they occupy the front position. Some bars, like the light beige ones, take on odd shapes. Overall, the graph is visually confusing and distracting, and thus hard to read, so it is best to avoid overlapping bars. The only exception that I recommend is when you completely overlap bars such that one set fits inside another in order to visually highlight an intimate relationship between them, as in the *correlation bar graph* that I introduced in Chapter 5, *Fundamental Variations in Graphs*. Here it is again:

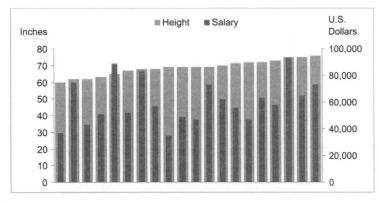

FIGURE 10.24 This is an example of fully overlapping bars, which illustrates an exception to the rule of avoiding overlap.

FILLS

The use of color or pattern to fill a bar should follow the general design practices that we've already examined:

- Avoid the use of fill patterns (e.g., horizontal, vertical, or diagonal lines), because they create disorienting visual effects.
- Use fill colors that are clearly distinct.
- Use fill colors that are fairly balanced in intensity for data sets that are equal in importance.
- Use fill colors that are more intense than others when you wish to highlight particular values above the others.

This last practice is a very useful way to direct your readers to pay particular attention to an important set of data, as in the following example:

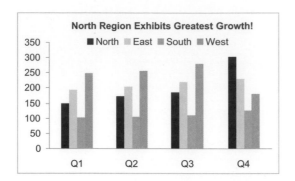

FIGURE 10.25 This is an example of using an intense fill color, in this case black, to highlight a particular set of values.

BORDERS

A border around a bar is only visible if the border's color is different from the fill color of the bar. The use of bar borders usually adds a visual component to the graph without adding information. Here are examples of some of the possible variations:

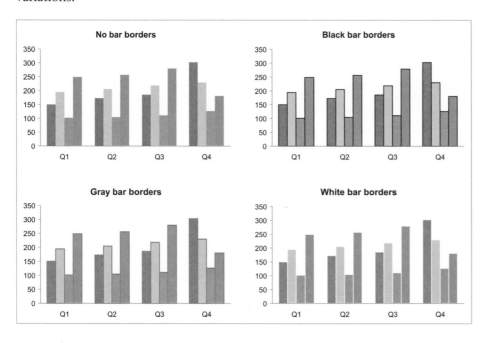

FIGURE 10.26 These examples show variations in the use of borders around bars.

Other than for highlighting, borders around bars are only useful when the fill color does not sufficiently stand out against the background (e.g., light yellow or blue bars against a white background). If you must use light-colored bars for some reason, the use of subtle borders (e.g., gray rather than black) creates the separation between the bar and the background that is needed to make the bars stand out.

Just like fill colors, borders may be intentionally introduced to make particular values or sets of values stand out from the others. In the following example, attention is clearly drawn to a particular bar through the use of a border that is absent from the other bars:

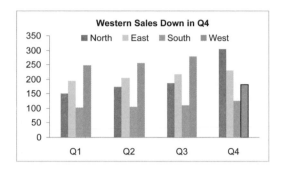

FIGURE 10.27 This example uses a border around a bar to highlight a particular value.

BASES

A bar consists of two ends: the one that marks the value, called the *endpoint*, and the one that forms the beginning, called the *base*. The purpose of this section is to consider where bars should begin. For all purposes but one, which we'll consider in a moment, bars should begin at zero and extend from there. This is not the same as saying that bars should always begin at the bottom or left edge of the graph. The axis that the bars rest on does not intersect the other axis at its lowest quantitative value when both positive and negative values are included. In the following example, the X axis intersects the Y axis near the middle:

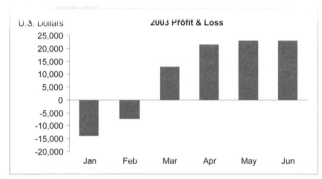

FIGURE 10.28 This is an example of a graph with an X axis that is not positioned at the end of the Y axis.

Whether the bars extend upward or downward, to the right or to the left, they should start at zero. Otherwise, their lengths do not correspond to the values they encode.

The only time a bar need not start at zero is when its purpose is to encode a range of values rather than a single value. Here's a graph that displays the range of employees' salaries by department, rather than a single value (e.g., an average) per department:

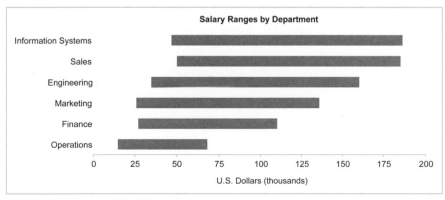

FIGURE 10.29 This is an example of bars that display ranges of values, rather than single values.

Bars of this type are called *range bars*, *high-low bars*, or *floating bars*.

Information that is encoded in a range bar can be enhanced by adding a significant value that falls within the range. A typical example is a range bar that displays the value of stock on a particular day, including its high, its low, and its closing values. Outside the realm of stock information, it is often useful to include measures of average in addition to the minimum and maximum values of a range. Here's an example that does just that:

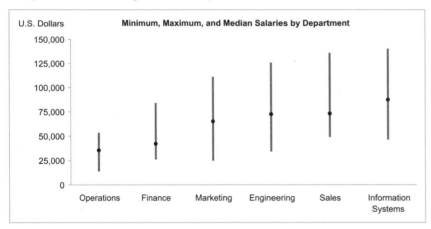

FIGURE 10.30 This is an example of range bars that include a point to mark an intermediate measure, in this case the median.

In this example, the ranges are encoded as vertical bars, and the medians are encoded as points. The bars are very thin, but they certainly don't need to be. Many software products that can be used to produce graphs do not offer a type of graph that is specifically designed for showing ranges. In order to illustrate a simple trick to get around this limitation, I used *Microsoft Excel* to create this example and selected a *Stock* graph to produce this display of high, low, and median values.

The bar design practices that we've covered may be summarized as follows:

Topic	Practices

Orientation	• Horizontal is the best *orientation* for bars when either of the following conditions exist:
	   • The graph displays a ranking relationship in descending order.
	   • The categorical subdivisions that label the bars won't fit side by side.
	• Avoid orienting text any way other than horizontally whenever possible. When there is no space for horizontal text, orient text at an incline, rather than vertically, whenever possible.

Proximity	• Set the width of white space separating bars that are labeled along the axis equal to 100% the width of the bars, plus or minus 50%. • Do not include white space between contiguous bars that are not labeled along the axis. • Do not overlap bars.

Topic	Practices
Fills	• Avoid the use of fill patterns (e.g., horizontal, vertical, or diagonal lines) because they result in disorienting visual effects. • Use fill colors that are clearly distinct. • Use fill colors that are fairly balanced in intensity for data sets that are equal in importance. • Use fill colors that are more intense than others when you wish to highlight particular values above the others.

Borders	• Only place borders around bars when one of the two following conditions exists:
	   • The fill color of the bars is not distinct against its background, in which case you can use a subtle border (e.g., gray).
	   • You wish to highlight one or more bars compared to the rest.

Bases	• Always start bars at zero except when each encodes a range of values rather than a single value.

Lines

Lines that are used to encode values in graphs fall into four categories:

- *Standard line*, which connects the individual values of a single data set
- *High-low line*, which displays a range of values across multiple data sets
- *Trend line*, which displays the overall trend of the values in a data set
- *Reference line*, which displays a set of values against which others can be compared

STANDARD LINES

In contrast to bars, which emphasize the distinctness of individual values, lines emphasize continuity and flow from one value to the next. Therefore, lines are particularly good at displaying values that change through time as well as the overall shape of that change.

When a graph contains multiple sets of quantitative values that are encoded as lines, you must use care to make them visually distinct. Lines that look too much alike are hard to trace as they cross one another, as in the following example:

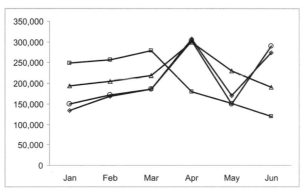

FIGURE 10.31 This graph shows lines that are not clearly distinct from one another.

In this example, the only visual characteristics that distinguish the lines are the distinct shapes of the points. This is not enough to enable easy distinction, especially where lines intersect. Shapes that mark individual data points don't solve this problem because they only appear intermittently along the length of the line. Unless you want to simultaneously display the overall shape of the data as well as the individual values, you wouldn't bother to mark the individual data points with shapes anyway, so you definitely shouldn't get into the habit of distinguishing lines by symbol shapes alone. Given what you know about the attributes of visual perception, how could you make the lines sufficiently distinct?

. . . .

The most effective method involves distinct expressions of color, either hue or intensity. Here are two new versions of the same graph without distinct point shapes but with distinct colors instead:

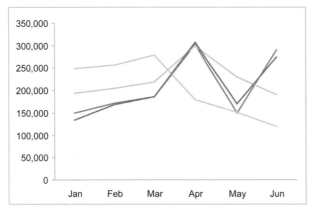

FIGURE 10.32 This graph shows hues that are used to make lines clearly distinct.

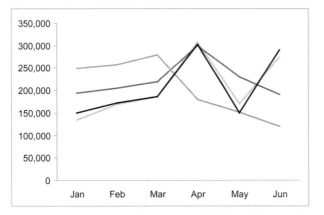

FIGURE 10.33 This graph shows different intensities of color that are used to make lines clearly distinct.

Hues work more effectively than differing color intensities, but if the graph is printed or photocopied without color, you can be confident that differing shades of gray will remain distinct.

The other attribute that is available for distinguishing lines, which is also useful when hues can't be used, is varations of line style, as in the following example:

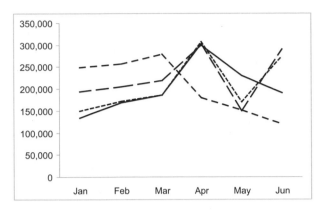

FIGURE 10.34 This graph shows line styles that are used in an attempt to make lines clearly distinct.

One drawback you can see here is that styles other than a solid line break up the flow of the line, which is somewhat distracting, and the appearance is jagged and therefore not as inviting to the eye. So avoid varying line styles unless no other attributes are available.

HIGH-LOW LINES

High-low lines connect the maximum and minimum quantitative values across multiple data sets at each location along a categorical scale. Here's an example:

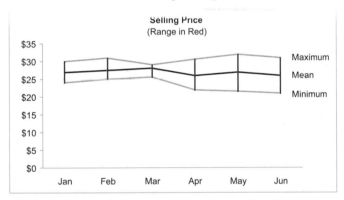

FIGURE 10.35 This is an example of a line graph with high-low lines.

As you can see, high-low lines can be used to display the range of values in multiple data sets. High-low lines can accompany data values that are encoded as lines, lines and points, or points alone. They may even be used without points or standard lines, as in the following example:

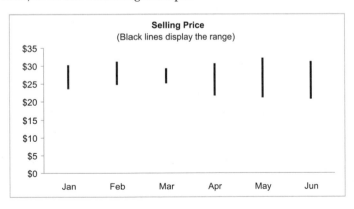

FIGURE 10.36 This is an example of high-low lines alone, without any other data values.

You may have noticed that high-low lines look very much like range bars. They look the same but sometimes vary in how they are constructed when using graphing software. Here are two more useful variations:

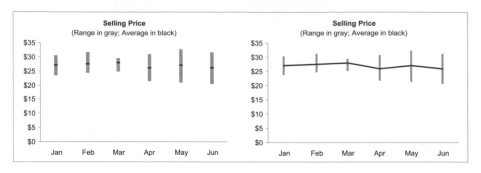

FIGURE 10.37 These examples show high-low lines, as well as an intermediate measure, displayed as points (on the left) and a line (on the right).

TREND LINES

Trend lines display the overall course of quantitative values that are spread across a series of categorical subdivisions. Trend lines are useful when displaying time-series relationships or correlations. Here's a time-series example:

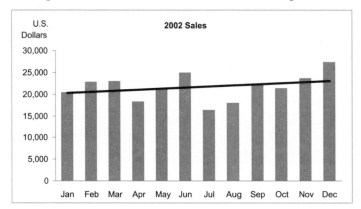

FIGURE 10.38 This graph shows a trend line that represents the overall course of sales across an entire year.

When trend lines are used with times-series relationships, the actual values don't need to be encoded as bars as they are in this example. The combination of bars and a trend line, however, does a nice job of combining a view of the individual values with a display of the overall direction of the data.

Here's another example, this time using a trend line to highlight a correlation in a scatter plot:

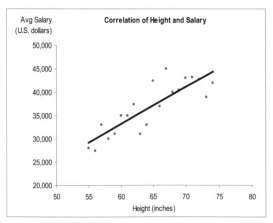

FIGURE 10.39 This graph includes a trend line that highlights a correlation in a scatter plot.

Trend lines do a great job of revealing the forest when you might otherwise get lost in the trees. Especially when you're dealing with a large number of erratic values, trend lines reveal what would be hard to see otherwise.

The primary design practice you should keep in mind regarding trend lines is the use of visual attributes to either highlight them or subdue them, as appropriate. This can be accomplished by using the attributes of color, line pattern, and line thickness that you now know quite well.

REFERENCE LINES

Reference lines are used to display a set of values against which others can be compared or to mark a point of interest along a categorical scale. Here's a simple example:

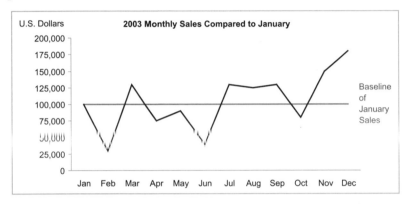

FIGURE 10.40 This graph shows a reference line that is used to mark the value of January sales as a reference to which the other months' sales can be compared.

This example illustrates a simple technique for showing how a set of values spread across a categorical scale compare to a specific value.

Quantitative reference lines are especially useful for displaying a measure of the norm, making it easy to see how values deviate from that norm. Averages (e.g., the mean and median) and standard deviations work well in this context. Here's an example of the same monthly sales values as in the graph above, but this time a reference line is used to compare them to average monthly sales:

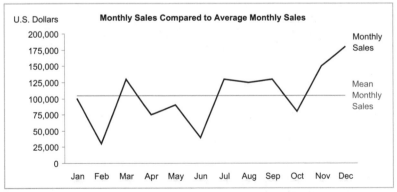

FIGURE 10.41 This graph shows a reference line that represents mean monthly sales as a reference to which each month's sales can be compared.

The next example is similar, but this time three reference lines are included to provide additional measures of the norm: one to display the mean, and two to display one standard deviation above and below the mean:

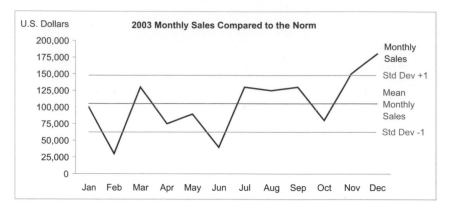

FIGURE 10.42 This graph shows reference lines that represent multiple measures of normal.

When reference lines are used to display measures of the norm in a scatter plot, the reference lines combine to form a rectangular area that defines the norm along both quantitative scales:

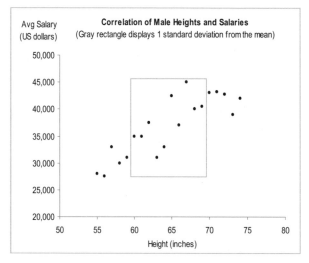

FIGURE 10.43 This scatter plot includes reference lines in the form of a rectangle that marks a measure of the norm, in this case one standard deviation above and below the means of heights and salaries.

When frames of reference are appropriate, like the rectangle in the example above, fill color may also be used to make the area within the frame stand out more.

Reference lines that mark points along a categorical scale are generally used to highlight a meaningful break in the values to form groups or to highlight an event on a time scale. The reference line in the following example divides subdivisions on the categorical scale of the Y axis into two groups:

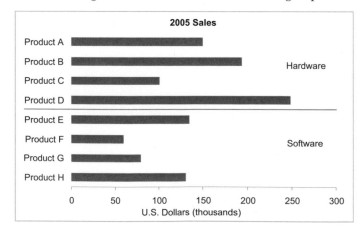

FIGURE 10.44 This graph shows a reference line that is used to mark a break point in a series of categorical subdivisions, in this case dividing products into two groups: hardware and software.

The reference lines in the following example mark events that are significant to the meaning of the data:

FIGURE 10.45 This graph shows reference lines that are used to identify significant events in time.

As this example illustrates, reference lines that mark events in time can be formatted as arrows with equal effect.

Scale Lines

Scale lines and *axes* are intimately related. A *scale*, regardless of the form it takes in the world at large (e.g., a ruler or a weight scale), measures the value of some thing. In graphs, scales are laid out along axes. They divide axes into increments of equal lengths and assign quantitative measures or categorical labels to those increments. Quantitative scales provide a means of assigning specific numeric values to the data objects that encode those values, based on their location along the scale line. Categorical scales assign categorical subdivisions to data objects in the same manner. The tiny lines that intersect scale lines to mark the increments are called *tick marks*. Without scale lines on the axes, graphs would be meaningless.

TYPES OF QUANTITATIVE SCALE

Quantitative scales come in a variety of types, but business graphs rarely require more than the following two:

- Common scale
- Logarithmic scale

In a *common scale*, the quantitative interval from one tick mark to the next is always the same. If the scale starts at 0, and the next tick mark as you go up the scale has a value of 10, then you know that the next tick mark after that will have a value of 20, and so on. To calculate the value of each tick mark, you simply add the value of a single interval to the value of the previous tick mark.

A *logarithmic* (a.k.a. *log*) *scale* works differently. Logarithms are shorthand for a series of numbers that are produced by raising the power of a *base* number for each interval by one. Here's an example of a logarithmic scale with a base of 10.

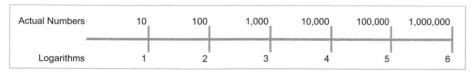

FIGURE 10.46 This illustration shows a log scale with a base of 10.

What's the base 10 logarithm of 10,000? The answer is 4, since 10^4 (i.e., 10 to a power of 4, or 10 * 10 * 10 * 10) equals 10,000. The standard way of writing this is $\log^{10}10{,}000 = 4$. Here's a concise definition:

> The logarithm of a number is the mathematical power to which
> another number, called the base, must be raised to equal that number.

In a logarithmic scale, then, actual values from one interval to the next are equal to the value of the previous interval multiplied by the base. In *Figure 10.46,* the actual value 1,000 is equal to the previous value 100 multiplied by the base of 10.

If you couldn't define it before, now you can, but what good is it? When are logarithmic scales useful in graphs? There are actually two circumstances in which logarithmic scales work better than common scales. Here's an example of the first and simplest circumstance in which a common scale exhibits a problem that a logarithmic scale can correct:

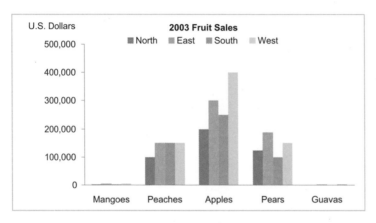

FIGURE 10.47 This is a graph that could be improved through the use of a logarithmic scale.

Do you see the problem? The huge difference between sales of apples at the high end and of guavas and mangoes at the low end makes it difficult to interpret the low values along this common scale. You could try to resolve this by making the graph much taller, but the result would look ridiculous. Logarithmic scales provide an easy solution to this problem. Here are the same values, this time distributed along a logarithmic scale:

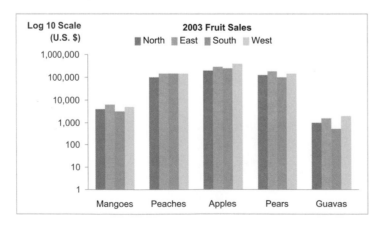

FIGURE 10.48 This graph uses a logarithmic scale to allow disparate values to be viewed together.

The huge gap between the sets of values has been dramatically reduced so that we can now easily examine all the data sets together. Notice, however, that great care has been taken in this example to alert the reader to the fact that a logarithmic scale has been used. This is important, because readers would minimize the difference in sales between guavas and apples if they failed to notice the logarithmic scale and adjust their perception accordingly.

Now for the second circumstance that benefits from a logarithmic scale. With a common scale, you know that when two pairs of bars differ in length by the same amount, they differ in value by the same amount. This is clearly not the case with logarithmic scales, but there is a unit of measure that will yield the same difference in value between equal differences in bar lengths along a logarithmic scale. Care to make a guess? Take a moment to think about it.

· · · · · · ·

The same distance anywhere along a logarithmic scale equals the same *percentage* or *ratio*. Here's the graph from the last example, but this time the equal distances between the sales in the north and sales in the east for each fruit have been highlighted:

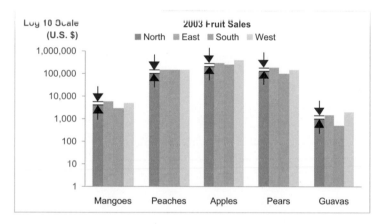

FIGURE 10.49 This example shows that, on logarithmic scales, equal distances correspond to equal percentages.

Sales in the east for every one of the fruits in this graph are precisely 50% greater than sales in the north. For instance, the comparison of apple sales between the north and the east is $200,000 to $300,000, and for apples and guavas it is $1,000 to $1,500; both equal a difference of 50%. This is something that you simply wouldn't notice using a common scale.

The logarithmic scale can be put to use, can't it? Whenever you want to compare differences in values as a ratio or percentage, logarithmic scales will do the job nicely. They are especially useful in time-series relationships if you wish to compare ratios of change across time. Let's take a look at some examples. In the first example on the next page, a common scale is used, which doesn't allow comparison of the rate of change from quarter to quarter across the four regions:

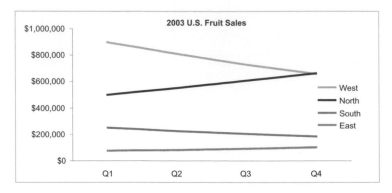

FIGURE 10.50 In this graph, a time-series relationship is displayed using a common scale. With a common scale, percentage change over time is difficult to discern.

In this next example, the same values are displayed using a logarithmic scale.

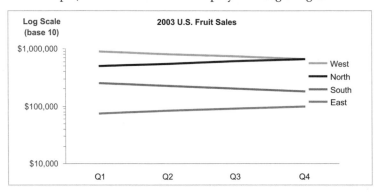

FIGURE 10.51 In this graph, the same time-series information that you find in *Figure 10.50* above is displayed using a logarithmic scale. Now percentage changes over time are easy to detect. Because both regions in which sales are decreasing have the same angle of descent, and both regions in which sales are increasing have the same angle of ascent, you know that their percentage changes are equal.

Notice that the descending angles of the west (beige line) and south (brown line) are the same, as are the ascending angles of the north (black line) and east (gray line). When you use lines to encode time-series values along a logarithmic scale, angles that are the same tell you that the rate of change is the same. In fact, the west and the south regions are both decreasing at a constant rate of 10% per quarter, and the north and east regions are both increasing at a constant rate of 10% per quarter. When the rate of change is the heart of your message (rather than change in actual quantity, such as actual dollars), a logarithmic scale is quite handy.

Despite the usefulness of logarithmic scales, the downside is that they are not visually intuitive. Equal distances along the scale correspond to equal logarithmic values but not equal common values. Note that the minor tick marks (i.e., the smaller ones that aren't labeled) along the logarithmic scale above are not equally spaced. Each minor tick mark represents an additional increment of the value represented by the labeled tick mark below it. The first minor tick mark above the one labeled $10,000 represents $20,000 (i.e., $10,000 + $10,000). The one above that represents $30,000, and so on up the scale until you reach $100,000; after that, each represents an additional $100,000. By displaying the minor tick marks, you can give your readers a visual clue that the scale is not common, but this isn't enough if they aren't accustomed to logarithmic scales. When logarithmic scales are new to your readers, take the time to provide a little education. You can head off inevitable confusion by including notes of explanation next to the graph where they can't be missed.

One more point to understand about logarithmic scales is that they don't have to use a base of 10. The base can be any number, but a base other than 10 is rarely seen in business graphs. In fact, software often restricts you to the use of base 10.

This is unfortunate, for logarithmic scales work in exactly the same way and offer exactly the same advantages with any base, and sometimes base 10 is too big to provide the best spread of values in a graph.

Base 2 is every bit as useful as base 10. With a base of 2, every major tick mark on the scale is exactly twice the actual value of the one before it (2, 4, 8, 16, 32, 64, 128, etc.). A base-2 logarithmic scale makes it easy to interpret rates of change in terms of doubling. A constant ascent up the scale from major tick mark to major tick mark for each increment of time indicates that the value has doubled during each period. The other advantage of a base-2 logarithmic scale is that if the values you are graphing are not spread across a broad range, this scale allows you to display greater detail than a log 10 scale that jumps from 10,000 to 100,000 to 1,000,000 in single bounds.

TICK MARKS

The information contained in a scale line consists solely of the tick marks, which establish the positions of values along the scale, and the labels, which assign numbers to the tick marks. The line on which the tick marks reside is simply an axis that plays a visual support role and carries no actual information. We'll explore the following questions:

- How visible should tick marks be?
- Where should tick marks appear on an axis?
- When can you eliminate tick marks?
- When should you use minor tick marks?
- How many tick marks should you use?
- Which values should you mark with tick marks?

How Visible Should Tick Marks Be?   Tick marks and labels allow us to interpret the data values in the graph and thus provide critical information, but they are not the values themselves. Consequently, they should be visually muted in comparison to the actual data values in the graph but prominent enough to be read easily.

Where Should Tick Marks Appear on an Axis?   You have three placement options: the inner side of the axis, the outer side of the axis, or across the axis. The following example illustrates these options:

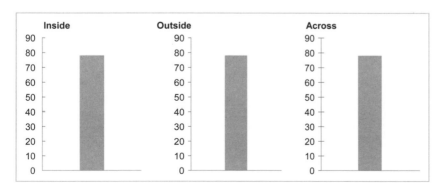

FIGURE 10.52: These examples show the three options for placing tick marks on an axis: inside, outside, and across.

As long as the tick marks are visually muted in relation to the data values, any of these options will work, but I prefer placing them on the outside, for this leaves the data region of the graph completely free of all but the values themselves, providing a clear backdrop for examining the values.

When Can You Eliminate Tick Marks?   Tick marks are always needed for quantitative scales, but what about categorical scales? Tick marks are superfluous on categorical scales. The categorical labels themselves identify the locations of the values clearly enough without assistance from tick marks. Take a look at the two pairs of examples below, one with categorical tick marks and one without:

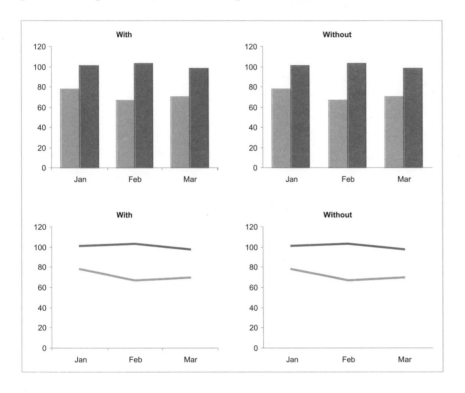

FIGURE 10.53: These are examples of graphs with and without tick marks on the categorical scale.

Even when there are many subdivisions along a categorical scale, tick marks serve no purpose. It is best to eliminate tick marks on a categorical scale.

When Should You Use Minor Tick Marks?   What about minor tick marks, the ones that can be included between the labeled tick marks to indicate finer increments along the scale line? Minor tick marks suggest a level of quantitative precision that graphs simply aren't meant to provide. They are only useful on logarithmic scales, not to provide greater precision, but simply to alert readers to the fact that a common scale isn't being used.

How Many Tick Marks Should You Use?   Another consideration is how many tick marks you ought to include on the scale line. You should always aim for a balance between so many that the scale looks cluttered and so few that your readers have difficulty determining the values of data objects that fall between them. Here are examples of both extremes:

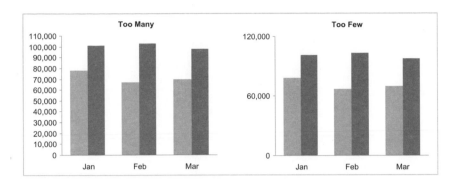

FIGURE 10.54: These are examples of graphs with too many and too few tick marks.

There is no exact number that works best in all circumstances, and the size of the graph is a factor that must be considered: the longer the scale line, the more tick marks it should contain. The following example shows a more appropriate number of tick marks, given the length of the scale:

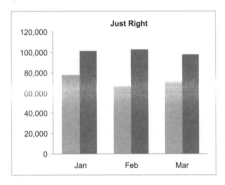

FIGURE 10.55: This graph shows an appropriate number of tick marks.

Which Values Should You Mark with Tick Marks?   Our final consideration regarding tick marks is the choice of values that they designate. It generally works best to use nice round numbers that your readers can relate to easily. Avoid values like the following:

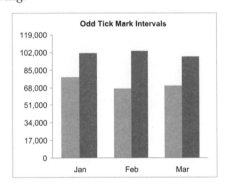

FIGURE 10.56 This graph shows tick marks that label values at odd increments, making them difficult to interpret.

If you use numbers like those above, you'll only lead your readers to wonder why the increments are so odd, searching for meaning where there is none.

Legends

Legends are the tiny tables that associate categorical subdivisions with the visual attributes that encode them (e.g., color or point shape). We'll look briefly at the following topics:

- When can you eliminate legends?
- Where should legends appear on the graph?
- How visible should legends be?
- Should legends have borders?
- How should you arrange labels in legends?

WHEN CAN YOU ELIMINATE LEGENDS?

Whenever categorical subdivisions are encoded in a graph, they must be labeled. If they appear on a categorical scale along an axis, they are labeled along the axis. However, if they are encoded using some other visual attribute, such as color, labels must be provided to tell your readers the meaning of the attribute. Legends are the conventional way to label these categorical subdivisions, but one circumstance lends itself to an alternate form of labeling. The graph below gives a hint. Take a look at it and see whether you can determine a different way to label the subdivisions that appear in the legend:

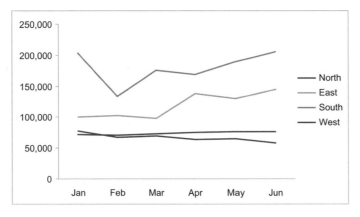

FIGURE 10.57 This graph includes a legend that could be replaced using a different means of labeling the categorical subdivisions of *North*, *East*, *South*, and *West*.

It's fairly easy to recognize in this example that the legend could be replaced by labels placed next to the lines themselves. The advantage is that the meaning of the various lines wouldn't need to be stored in short-term memory by those using the graph, making it easier and faster to read. Here's the improved version:

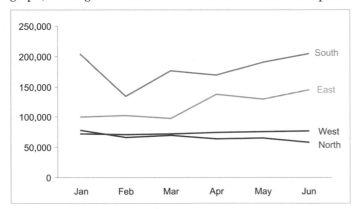

FIGURE 10.58 In this graph categorical subdivisions are labeled directly, next to the lines that encode them, rather than separately in a legend.

The answer to the question "When can you eliminate legends?" is "Whenever the data subdivisions that need labels are grouped together so that you can place a label right next to each set." When lines are used to encode sets of values, each line acts as a grouping mechanism, making it easy to place a label right next to the line. However, as you can see in the next example, bars don't work this way:

If you are producing graphs using software that doesn't offer the option of placing labels right next to the data sets, you can often type labels in yourself, which is what I did to produce *Figure 10.58* using *Microsoft Excel*.

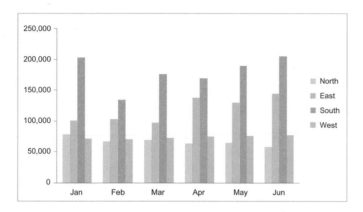

FIGURE 10.59 This is a graph that must use a legend to label the categorical subdivisions, because the subdivisions are encoded as bars.

You certainly wouldn't want to repeat the same labels over and over again next to each bar. Sometimes separate sets of data points in scatter plots are grouped together closely enough to label them right next to the points, but this is rare and should be avoided if the labels would disrupt perception of the pattern formed by the points.

WHERE SHOULD YOU PLACE LEGENDS?

Legends are generally placed outside of the data region, but this needn't be a rigid rule. If you can place the legend inside the data region without getting in the way of the data values or interfering with perception of their shape, there is no reason why you shouldn't do so. In fact, the closer the legend is to the data values, the easier it is to read the graph. Legends may be placed anywhere they fit and don't interfere with other more important components of a graph.

HOW VISIBLE SHOULD LEGENDS BE?

Legends provide the means to interpret categorical data, which is critical information, but they are not the actual data themselves. Therefore, although they should be clearly visible in the graph, they should be somewhat less prominent than the actual data. You want your readers' eyes to be drawn predominantly to the data region of the graph where the actual data reside.

SHOULD LEGENDS HAVE BORDERS?

You may have noticed that few of the legends that appear in examples throughout this book have borders around them. This is an intentional omission. Based on your understanding of visual perception, why do you think this is? Look at the two graphs below to see the difference:

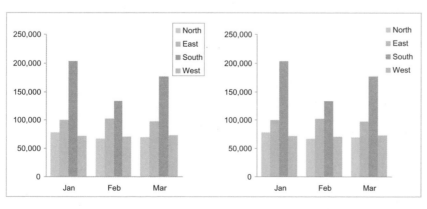

FIGURE 10.60 These two graphs are the same except that one includes a border around the legend and the other does not.

First, the border around the legend does not add any meaning to the graph. The legend doesn't need a border in order to be clearly distinct from other parts of the graph. Second, enclosing the legend with a border draws undesirable attention to it. The legend isn't where you want your readers' eyes to be drawn.

When a legend is placed near another graph component and these components must appear distinct from one another, something is needed to visually distinguish the legend from that other component; in this case, a border is useful. Be sure to make the border visually subtle, just visible enough to set the legend apart.

HOW SHOULD YOU ARRANGE LABELS IN LEGENDS?

You are probably accustomed to seeing the labels in legends arranged vertically rather than horizontally, but either arrangement can work fine. It really depends on the space you're working with and how the data is encoded. In the following example, the legend is arranged horizontally below the title so that none of the graph's width is sacrificed:

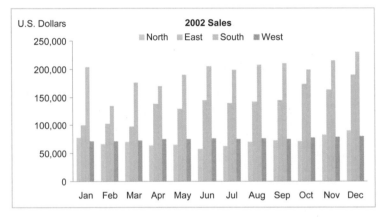

FIGURE 10.61 In this graph the labels in a legend are arranged horizontally.

Not only do the horizontal arrangement of the labels and placement of the legend above the data region save horizontal space in the graph, they also arrange the labels in the order of the corresponding bars, which improves the graph's ease of use.

Other Text

In addition to text that appears in legends and as labels along axes, text also appears in or adjacent to graphs as titles and notes. Regardless of its purpose or form, text should be located as close to the information that it complements as possible without interfering with that information. By convention, titles generally appear just above the data region of the graph, and notes generally appear just below the graph. As long as they don't disrupt perception of the visual information in the data region, notes may be placed in the data region, especially when they apply to particular data. Notes that are instructional or interpretive in nature, however, usually don't need to appear in the data region and can be placed adjacent to the graph (below, above, or to either side), close enough that they are not overlooked. Here's an example of effective text placement:

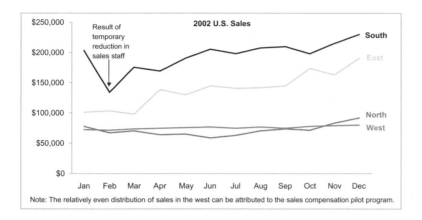

FIGURE 10.62 This graph demonstrates text placement near to the information that it complements without hindering perception of that information.

Limit to the Number of Categorical Subdivisions

There is no conceptual limit to the number of categorical subdivisions you can include in a single graph, but there is definitely a practical limit. Too many visually distinct categorical subdivisions cannot be decoded without great effort, because, as you learned in Chapter 6, *Visual Perception and Quantitative Communication*, we can only hold the meanings of a few distinctions in short-term memory at one time. We've all seen graphs like the following example:

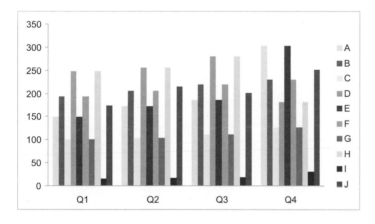

FIGURE 10.63 This graph has too many categorical subdivisions.

Not only is it impossible to keep in mind what each colored bar represents, the graph is cluttered. Clutter is visually exhausting, discouraging any effort to read the graph. The upper limit to the number of categorical subdivisions that can be effectively displayed in a graph is somewhere between five and eight. This is true whether you encode the data using bars, points, or lines.

Support Component Design

The only components in graphs that simply provide visual structure and don't represent information are the axes and the area defined by the axes where the data-encoding objects appear.

Axes and the Data Region

Axes give dimensionality to graphs. A single axis, either vertical or horizontal, produces a 1-D graph. Two axes—one vertical and one horizontal—produce a 2-D graph, which is by far the most common type. The space that is defined by the axes where the data values are plotted (i.e., where the points, lines, or bars reside) is called the *data region* or the *plot area*. When you design a graph, you need to ask three questions regarding the axes and data region:

- Should the graph include one, two, three, or four axis lines?
- What should the relative lengths of the axes be?
- What visual attributes should the data region have?

SHOULD THE GRAPH INCLUDE ONE, TWO, THREE, OR FOUR AXIS LINES?
2-D graphs usually include one vertical axis and one horizontal axis. There are alternatives, however, which include the following, in order of commonality:

- Two vertical (left and right) and two horizontal (bottom and top) axis lines
- One horizontal and no vertical axis line
- One vertical and no horizontal axis line

Here's an example of each variation, including the usual method:

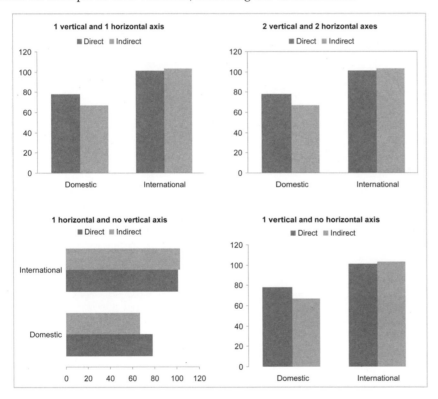

FIGURE 10.64 These graphs illustrate various axis configurations.

Under most circumstances, the standard combination of one vertical and one horizontal axis line works best. Two perpendicular lines, tracing only two sides of the rectangular data region, sufficiently define the space, except in rare circumstances. Including two pairs of vertical and two pairs of horizontal axes to form a complete rectangle around the data region is only necessary when the data region must be emphasized and separated from surrounding text and objects because those surroundings would otherwise compete too forcefully for attention.

An axis line may be left off without adverse affect when it hosts a categorical scale and the data values are encoded as horizontal bars. Because bars begin at the axis, their edges trace the line that the axis would otherwise display, which adequately delineates the data region. This works fine for horizontal bars, but I find that without a base, vertical bars appear to float in space without an appropriate visual foundation, as in the bottom right example on the previous page. Graphs with horizontal bars can omit the vertical axis line except when the data labels and other text along the left side of the graph need additional separation from text elsewhere in the graph.

WHAT SHOULD THE RELATIVE LENGTHS OF THE AXES BE?
The ratio of the length of the vertical axis to the length of the horizontal axis, or stated differently, the ratio of the data region's height to its width, is called the *aspect ratio*. It is calculated as height divided by width. The aspect ratio of a graph's data region greatly influences perception of the data. Here are a few examples of aspect ratios that vary from 0.5 on the low end to 2.0 on the high end:

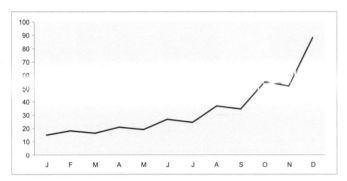

FIGURE 10.65 This graph has an aspect ratio of 1 to 2, or 0.5.

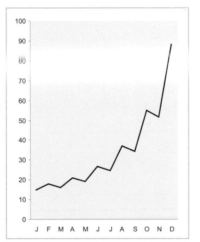

FIGURE 10.68 This graph has an aspect ratio of 1.5 to 1, or 1.5.

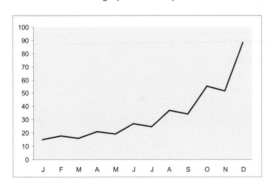

FIGURE 10.66 This graph has an aspect ratio of 1 to 1.5, or 0.67.

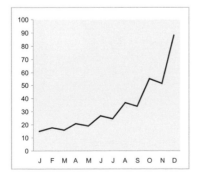

FIGURE 10.67 This graph has an aspect ratio of 1 to 1, or 1.0.

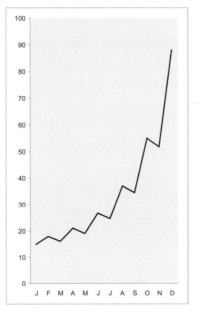

FIGURE 10.69 This graph has an aspect ratio of 2 to 1, or 2.0.

All of these graphs contain the same data; only their aspect ratios differ. When lines are used to encode values, as the aspect ratio increases, so does the appearance of the rate of change. A rate of change that already looks impressive with an aspect ratio of 0.5 looks like a blast-off at Cape Canaveral with an aspect ratio of 2.0. Both are accurate, both display precisely the same numbers, but they certainly differ in perceptual impact.

There is no single aspect ratio that is always best. There are two design practices, however, that you should keep in mind. The first is that you should not manipulate the aspect ratio to intentionally exaggerate or downplay the rate of change. If your graphs usually appear wider than they are tall, suddenly making one taller than it is wide to convince your readers that sales are going through the roof would qualify as manipulation.

The other general practice is to stick to the convention of making your graphs wider than they are tall. Emphasizing the horizontal rather than the vertical generally makes graphs a bit easier to read and more in line with what people are accustomed to seeing. Scatter plots are the primary exception to this practice, which usually display data best in a square-shaped data region.

WHAT VISUAL ATTRIBUTES SHOULD THE DATA REGION HAVE?

Apart from the aspect ratio of the data region, its only other visual attribute is its fill color. Once again, our objective is to make the data stand out above the other components of the graph. We want our readers' eyes to be drawn to the data. This is accomplished by making the data objects visually prominent, not by energizing the background with vibrant color. Unless your data objects are rendered as light colors, they will stand out best against a light background. A white data region is generally the best background for your data objects, but there are times when other light shades of color may be useful, such as light gray or yellow. When the data region of a graph needs a little extra help to stand out against its surroundings, a subtle fill color generally does the trick. Here are a few examples of fill colors that work:

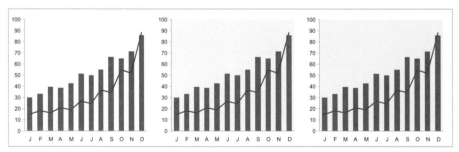

FIGURE 10.70 These examples show data region fill colors that provide an appropriate background for the data.

Another way to subtly highlight the data region is to use fill color for the graph except in the data region. By leaving the fill color of the data region white and coloring the surrounding areas of the graph with a light shade of color, the white of the data region stands out in contrast, as in the following examples:

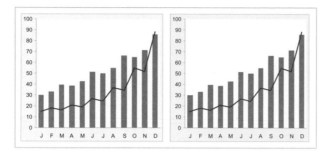

FIGURE 10.71 These graphs highlight the data region by using a subtle fill color for the areas that surround the data region.

When the data region needs an extra visual boost, the other visual attributes of the graph, especially the colors of the data objects, determine which of the two methods will work best (i.e., a fill color for the data region or for the surrounding areas of the graph). For instance, if your graph contains several sets of bars, and the color of one is light, the use of a subtle fill color in the data region might help them stand out more clearly than they would against white.

Grid Lines

From the early days of quantitative graphs to the advent of computer-generated versions, one of the primary purposes of grid lines was to make it easier to draw graphs. The lines served as guides, assisting the placement and sizing of components. Now that almost all graphs are generated using computers, this purpose is obsolete. Today, grid lines are used to enhance perception in three ways:

- Enhance the look-up of values
- Enhance the comparison of values
- Enhance perception and comparison of localized patterns

In all cases, grid lines are support components, providing assistance in the visual perception of the data values themselves. Thus, grid lines should always be visually subducd. Thin, light lines (e.g., light gray) work best. Dark or heavy grid lines, such as those in the example below, should always be avoided, because they disrupt perception of the data:

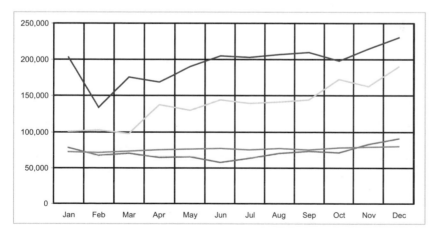

FIGURE 10.72 This graph includes dark, heavy grid lines, which you should always avoid.

ENHANCE THE LOOK-UP OF VALUES

The use of grid lines to assist readers in looking up values is probably the most common but least necessary. We've established that tables work better for looking up precise values. Because graphs are used primarily to present the shape of the data, precise values aren't necessary. Nevertheless, grid lines do help when the tick marks alone do not provide the level of precision that is needed when reading data values in graphs. This situation is most common with graphs that are especially wide, which makes it difficult to assign values to data objects that reside far from the scale line. In the example below, subtle grid lines provide all the look-up assistance that is necessary:

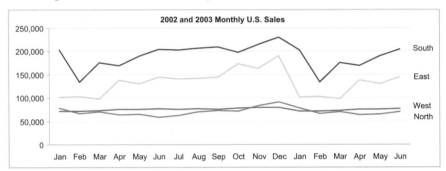

FIGURE 10.73 This graph uses subtle grid lines to enhance look-up.

ENHANCE THE COMPARISON OF VALUES

We learned in the earlier chapter on visual perception that we perceive differences in the sizes of objects, such as the length of bars, as percentage differences rather than absolute differences. Two long lines that differ by only a centimeter in length appear to be the same whereas two short lines with respective lengths of only two centimeters and one centimeter appear quite different. When perception of subtle differences is necessary, grid lines help by dividing the data region into smaller units, which make otherwise subtle differences in size and location obvious. Grid lines allow us to compare differences in the size and position of objects relative to a much smaller space than the entire data region of the graph, which significantly increases the percentage differences, making these differences easier to see. The following examples illustrate this point:

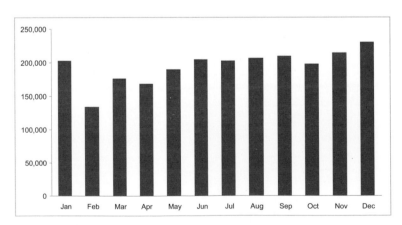

FIGURE 10.74 This graph demonstrates how difficult it is to perceive differences in the size and location of data objects when those differences are relatively small.

It's difficult to assess differences between bars that are close in length, such as *Jan*, *Jun*, *Jul*, and *Aug*. However, with the addition of subtle horizontal grid lines in the example below, these differences become much easier to discern:

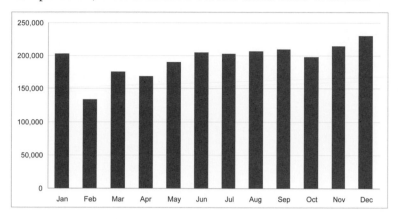

FIGURE 10.75 This graph uses grid lines to make it easier to perceive differences in the lengths of the bars.

ENHANCE PERCEPTION AND COMPARISON OF LOCALIZED PATTERNS

Graphs do a marvelous job of revealing the overall shape of an entire collection of values. If you need to enhance perception of the shape of values in a sub-section of the graph, however, grid lines do the job nicely. This perception is enabled by the same visual mechanism that enhances our perception of differences. This works best in scatter plots that use subtle grid lines to help the reader focus on a subsection of the data region. Here's a typical scatter plot:

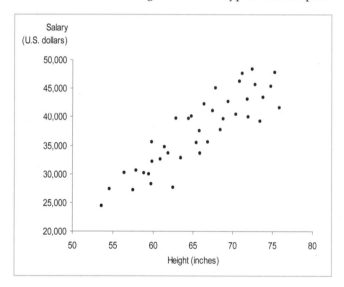

FIGURE 10.76 This graph would be difficult to read if you needed to isolate subsets of data.

This scatter plot only contains 40 individual data points, which is a small number compared to the 625 data points that we can detect in a square inch of space. Nevertheless, it would be difficult to focus on the pattern formed by the subset of values that fall in the range of 65 to 70 inches and $35,000 to $40,000. However, the addition of subtle grid lines makes this task easy, as shown in the next example:

This quantity of 625 data points was taken from Edward R. Tufte (1983) *The Visual Display of Quantitative Information*. Cheshire CT: Graphics Press, page 161.

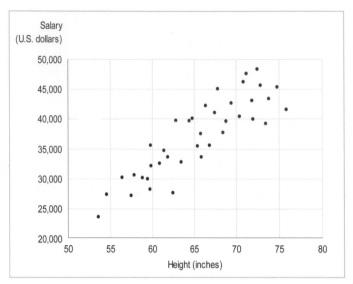

FIGURE 10.77 This graph uses subtle grid lines to enhance our ability to perceive specific subsets of data.

These grid lines need not correspond to tick marks along the scale line. As long as they divide the space evenly, they need not correspond to any particular values along the scale. For instance, it is sometimes useful to simply divide a scatter plot into quadrants, as in the following example:

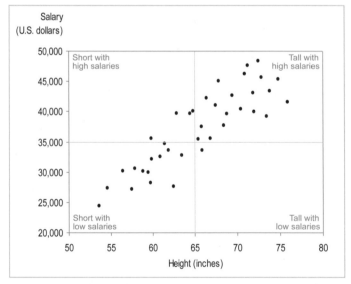

FIGURE 10.78 This scatter plot uses two intersecting grid lines to evenly divide the data region into quadrants.

The use of grid lines to divide scatter plots into quadrants can bring features of the data to light that might otherwise be missed. This example clearly reveals, with its empty lower right quadrant, that none of the shorter employees earn high salaries.

When we use a series of related graphs that contain similar data and the same scales, grids help us to compare the same regions along those scales in multiple graphs. Let's say that we collected the height versus salary data separately for 10 different geographical regions and wanted to not only look for overall regional differences but also for localized differences within particular ranges of values across regions. Grids make it easy to compare the same rectangular region in multiple graphs, enabling us to detect subtle changes in the data. Here's an example of two sets of height versus salary data: one for the west and one for the east. Look at how easy it is to determine the differences in the two regions that are located in the range of 70 to 75 inches and $45,000 to $55,000:

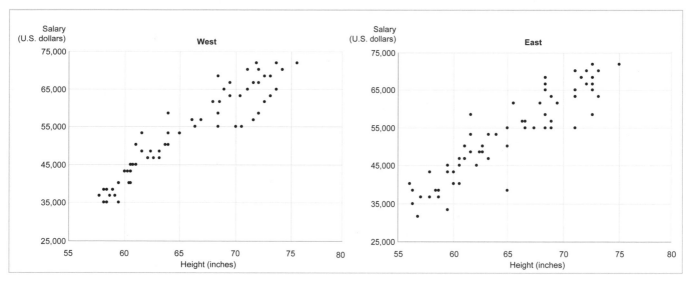

The use of multiple graphs that contain similar data is a powerful means of making comparisons, which we will explore in the next chapter.

FIGURE 10.79 This example demonstrates the use of grid lines to enable the comparison of the same subregions in two graphs.

Summary at a Glance

Component	Practices

Points	• When sets of points cannot be clearly distinguished, correct by either:
	  • Enlarging the points
	  • Selecting objects that are more visually distinct
	• When points overlap such that some are obscured, correct by:
	  • Enlarging the graph and/or reducing the size of the points
	  • Removing the fill colors
	  • Selecting radically distinct shapes, such as circles and crosses
	• When points are obscured by lines, correct this by making the points comparatively prominent.

Bars	• Use horizontal bars when either of the following conditions exist:
	  • The graph displays a ranking relationship in descending order.
	  • The categorical subdivisions that label the bars won't fit side by side.
	• Proximity
	  • Set the width of white space separating bars that are labeled along the axis equal to the width of the bars, plus or minus 50%.
	  • Do not include white space between bars that are not labeled along the axis.
	  • Do not overlap bars.

Component	Practices

Bars *(continued)*	• Fills
	  • Avoid the use of fill patterns (e.g., horizontal, vertical, or diagonal lines).
	  • Use fill colors that are clearly distinct.
	  • Use fill colors that are fairly balanced in intensity for data sets that are equal in importance.
	  • Use fill colors that are more intense than others to highlight particular values.
	• Only place borders around bars when one of the two following conditions exists:
	  • The fill color of the bars is not distinct against its background, in which case you can use a subtle border (e.g., gray).
	  • You wish to highlight one or more bars compared to the rest.
	• Always start bars at zero except when each encodes a range of values.

Log Scales	• Use to reduce the visual difference between quantitative data sets with significantly different ranges of value so they can be clearly displayed together • Use to compare differences in value as percentages
Tick Marks	• Mute them in comparison to the data objects. • Place them outside the data region. • Use them with quantitative scales but not with categorical scales. • Aim for a balance between including so many that the scale looks cluttered and using so few that your readers have difficulty determining the values of data objects that fall between them. • Avoid using them to mark values at odd intervals.
Legends	• Use them for categorical labels when the labels are not associated with a categorical scale along an axis and cannot be directly associated with the data objects. • Place them as close as possible to objects they label without interfering with other data. • Render them less prominent than the data objects they label. • Use borders around legends only when necessary to separate legends from other information.
Axes	• Don't manipulate the aspect ratio to distort perception of the values.

Grid Lines	• Thin, light grid lines may be used in graphs for the following purposes:
	  • Enhance look-up of values
	  • Enhance comparison of values
	  • Enhance perception of localized patterns

Miscellaneous	• Place titles and notes as close to the data values that they complement as possible without interfering with those values. • Include no more than five to eight data sets in a single graph.

11 DESIGN SOLUTIONS FOR MULTIPLE VARIABLES

Graphs can be used to tell complex stories. When designed well, graphs can combine a host of data spread across multiple variables to make a complex message accessible. When designed poorly, graphs can bury even a simple message in a cloud of visual confusion. Excellent graph design is much like excellent cooking. With a clear vision of the end result and an intimate knowledge of the ingredients, you can create a whole that nourishes and inspires.

Combining multiple units of measure
Combining multiple graphs in a series
    Consistency
    Arrangement
    Sequence
    Rules and grid lines

Often, we need to communicate messages that involve more than one quantitative variable grouped in relation to several categorical variables. Sometimes, to communicate effectively we need to help our readers see the information from more than one perspective. A single graph can often be used to tell a complex story elegantly, but sometimes a single graph simply won't do. This chapter focuses on design strategies that address the presentation of multiple variables in two useful ways: combining multiple units of measure in a single graph and combining multiple graphs in a series.

Combining Multiple Units of Measure

It is easy to combine multiple sets of quantitative data in a single graph when they all use the same unit of measure. For instance, in a single time-series graph, you could easily display revenue, expenses, and profits, if they all use U.S. dollars as their unit of measure. In other words, they share the same quantitative scale, as in the following example:

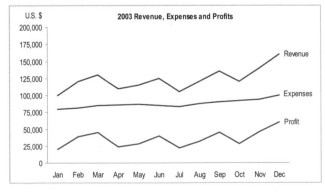

FIGURE 11.1 This graph displays multiple sets of quantitative data, all sharing the same unit of measure.

But what do you do to combine two related sets of quantitative data so they can be compared to one another when they have different units of measure? A typical example involves time-series sales information consisting of revenue in dollars and order volume as a count. One solution is to create two graphs and place them close to one another in a manner that makes comparisons easy. This is often the best strategy, and we'll explore it in detail later in this chapter. There is a way, however, to combine these two units of measure in a single graph. In the example below, you see the information presented as two graphs:

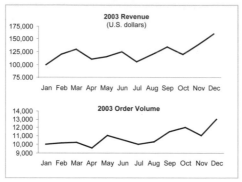

Think for a moment about how these two graphs, with their different scales, might be combined into a single graph.

· · · · · · ·

The solution involves taking advantage of the fact that a single dimension in a graph, either its height or width, has two sides. Because the graphs in our example above involve time-series relationships, we know that the horizontal axis must provide the categorical scale of time and that the vertical axis must contain a quantitative scale. The vertical dimension of the graph can accommodate two scale lines: one on the left with one unit of measure and another on the right with the other. Here's how it might look:

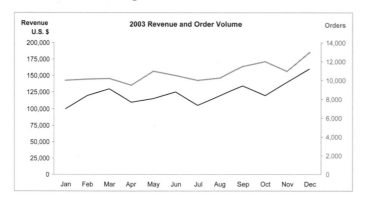

Now the two related measures can be compared quite easily even though they use different units of measure. Perhaps too easily, however. Even though each line is associated with a different scale, by displaying them in a single graph, we are tempting people to do something they shouldn't: to compare the magnitudes of values on one line to those on the other. Because their scales are different, this comparison is meaningless.

You can often enhance quantitative messages by displaying the same data from multiple perspectives. When you can't combine all of the data in a single graph, it is worthwhile to create multiple graphs with varying perspectives. Examining something from several angles often elicits insights that might have been missed entirely when viewing it only from one. Your message will often speak most clearly and persuasively when it is constructed from a combination of graphs, tables, and text, collaborating as a team. There is probably no better or more common example of combined displays of quantitative communication than the partnership between one or more graphs to show the overview and the shape of the data with one or more tables to provide the supporting details.

Combining Multiple Graphs in a Series

You can only squeeze so many sets of data into a single graph. It is this limitation of bar graphs that often tempts people to add a 3-D graph, with disappointing results. However, there is a solution that extends by one the number of data sets that can be displayed, which is often enough. This is the use of multiple graphs arranged together as a series.

Let's walk through a typical scenario, beginning with the need to display bookings and billings by sales region for last month. Here's what you have so far:

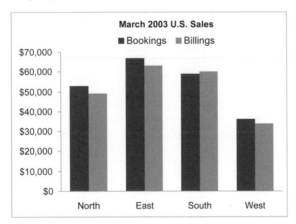

FIGURE 11.4 This graph displays two sets of quantitative values (i.e., bookings and billings) and one set of categorical subdivisions (i.e., sales regions).

Simple, so far. Now, if you need to add another set of related quantitative values (e.g., profits), you could do so easily by adding another set of bars. But what if you need to show sales not only by sales regions but also by sales channels (e.g., direct, distributor, and reseller sales)? This requires the addition of another set of categorical subdivisions. Adding a third axis would result in utter visual confusion, so this option is out. What can you do?

The answer involves multiple graphs arranged closely together so they can be examined together. Here's a simple example that solves our problem:

FIGURE 11.5 This is a series of related graphs that display two sets of quantitative values (i.e., bookings and billings) by two sets of categorical subdivisions (i.e., sales regions and sales channels).

Even though this involves three separate graphs, the nature of the arrangement allows them to be examined as one super-graph with three sets of axes. This series of graphs happens to consist of three because there are only three sales channels, but it could consist of as many graphs as you could fit together within the span of the eye. As the number of graphs grows, the trick is to reduce their individual size enough to allow them to be seen together on a single page or screen. You can arrange the graphs horizontally (i.e., side by side), vertically (i.e., one on top of the other), and at times you can do both, producing a matrix of graphs arranged in multiple columns and rows.

Graphs in a series like this should contain the same data, varying only by a single set of categorical subdivisions, as in the example above, which varies only by sales channel. Because everything about the graphs is the same except this one variable, they are very easy to view together and compare. Edward Tufte refers to such an arrangement of related graphs as *small multiples*. Tufte explains that "Small multiples resemble the frames of a movie: a series of graphics, showing the same combination of variables, indexed by changes in another variable."[1]

To learn to effectively design related graphs in a series, we'll examine the following topics:

- Consistency
- Arrangement
- Sequence
- Grid lines

1. Edward R. Tufte (1983) *The Visual Display of Quantitative Information.* Cheshire CT: Graphics Press, page 170.

Consistency

Most important when designing a series of related graphs is the maintenance of consistency among the graphs. Consistency is required for comparison. Consistency also contributes a great deal to efficient interpretation. Your readers only need to learn how the first graph works and can then quickly apply that knowledge to each graph in the entire series because the graphs all work in precisely the same way.

When you design a related series of graphs, make sure that every visual characteristic of each graph is the same. This includes the aspect ratio of the axes, the colors used to encode the data, the font used for text, and so on. Any difference will slow your readers down and induce them to search for meaning in that difference.

Pay particular attention to the scales along both of the axes. Many software products that generate graphs automatically adjust quantitative scales to fit the range of values. When graphs are combined in series, this is a significant problem, because the graphs can only be compared accurately if their scales are exactly the same. Look at what happens to our previous example when we allow the software to adjust the quantitative scale:

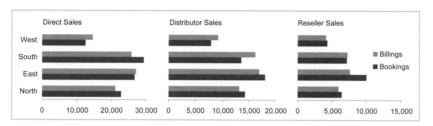

FIGURE 11.6 This example demonstrates the problem that results when quantitative scales vary in a series of related graphs.

As you can see, the differences in sales between the three sales channels have been visually reduced because of the variations in scale.

Make sure that the categorical scale also remains consistent with subdivisions in the same order (e.g., always *West*, *South*, *East*, and *North*) and the same full set of subdivisions in each graph, even when a value is zero or null. Here's the same series again, but this time there are no distributor sales in the south. Rather than displaying a value of zero, the south region was excluded from the graph entirely, resulting in a confusing inconsistency:

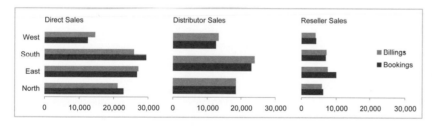

FIGURE 11.7 This series of related graphs exhibits an inconsistency in the sets of sales region subdivisions: the south region is missing in the graph of distributor sales.

One type of component in these graphs does not need to remain consistent throughout. Take a look at our original example again (*Figure 11.5* on the left-hand page) to find the components that are not the same in each graph, not counting the differing titles. Notice that the text that labels the four sales regions only appears in the left-most graph, and the legend that labels billings and bookings only appears in the right-most graph. Because the graphs are contiguous and consistent, these labels don't need to be repeated in each graph. In fact, to repeat them would result in unnecessary data ink.

Arrangement

The best arrangement for a series of graphs, whether horizontally side by side, vertically one on top of the other, or both in the form of a matrix, depends primarily on your answer to the question "Which set of categorical subdivisions do you want to make it easiest for your readers to compare?" Graphs generally include only one quantitative scale except when they include two to display a correlation relationship in a scatter plot. With the exception of scatter plots, then, graphs that are arranged in a series of small multiples include one set of categorical subdivisions that varies from graph to graph and one set that provides a categorical scale along either the horizontal or vertical axis. Other sets of categorical subdivisions may also be included that aren't associated with an axis, but these aren't relevant to the issue at hand. Of the two sets of categorical subdivisions (i.e., the one that varies from graph to graph and the one aligned with an axis), which subdivisions do you want most to help your readers compare?

If it is the one that varies from graph to graph, then you should arrange the graphs so that their quantitative scales are aligned for easy comparison. If their quantitative scales are attached to their vertical axes, then you should arrange the graphs horizontally (i.e., side by side). If their quantitative scales are associated with their horizontal axes and the values are encoded as horizontal bars, then you should arrange the graphs vertically (i.e., one on top of the other). In the example below, the quantitative scales run along the horizontal axes, so the graphs have been arranged vertically.

FIGURE 11.8 This series of graphs is arranged for easy comparison of the categorical subdivisions that vary from graph to graph, in this case the three sales channels (i.e., *Direct Sales*, *Distributor Sales*, and *Reseller Sales*).

Notice how easy it is to compare the three sales channels: *Direct Sales*, *Distributor Sales*, and *Reseller Sales*.

If, on the contrary, you wish to make it easiest for your readers to compare the categorical subdivisions associated with the vertical axes in these same graphs, which in this case is the set of sales regions (*North*, *South*, *East*, and *West*), you would want to arrange the graphs horizontally, as in the example below:

FIGURE 11.9 This series of graphs is arranged for easy comparison of the categorical subdivisions that are associated with the vertical axes (i.e., *West*, *South*, *East*, and *North*).

This arrangement makes it easier to compare the performance of a single sales region, such as the south, simply by scanning across from one graph to the next. Comparing sales in the south across the three graphs would be more difficult if the graphs were arranged vertically, as they were in *Figure 11.8*. Take a moment to see for yourself.

Because graphs that display correlations include quantitative scales on both axes, they may be arranged horizontally, vertically, or in both directions as a matrix. Arranging graphs as a matrix allows you to squeeze several graphs onto a single page or screen. Here's an example that includes six graphs:

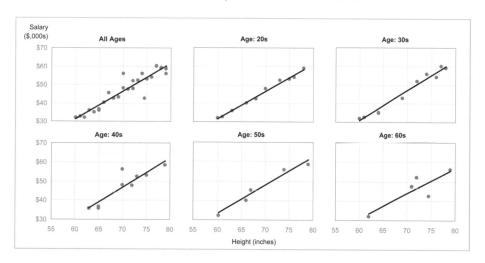

FIGURE 11.10 This is a series of scatter plots arranged as a matrix. Employees' heights and salaries are correlated along the X and Y axes, and their ages are grouped into separate graphs.

As you can see, many more than six of these graphs could have been squeezed onto a single page or screen.

Any type of graph can be arranged in a matrix of small multiples (time-series, part-to-whole, etc.), but this arrangement excels for the display of scatter plots, which can incorporate an extremely large amount of data for simultaneous viewing and comparative pattern detection. This allows you to display correlations between three sets of variables: two aligned with the X and Y axes of each graph and a third varying with each graph in the larger matrix. Even when the graphs include a categorical scale, if you can't fit the entire series across or down a single page or screen, a matrix arrangement is the best alternative. For instance,

having to scan one at a time down three columns of time-series graphs on a
single page or screen is a negligible disadvantage compared to the perceptual loss
that would result if they were spread across three separate pages or screens.
Notice how nicely this works:

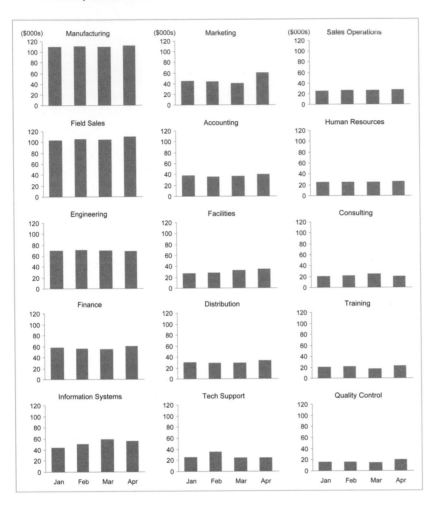

FIGURE 11.11 This series of graphs
has been arranged in a matrix, even
though it includes a categorical axis
(i.e., *Jan*, *Feb*, *Mar*, and *Apr*).

Always do your best to keep the entire series within eyespan.

A single series of related graphs like those we've examined all share the same
categorical and quantitative scales. The same one or more measures appear in
each graph, such as booking and billing dollars, order count, expense dollars, or
headcount. It is often useful to compare multiple units of measure, but the
technique that we examined earlier in the *Combining Multiple Units of Measure*
section is limited to two, with each on an opposite side of the graph. The
arrangement of graphs in a matrix, however, allows us to escape this limit.

Imagine that you want to display forecast versus actual revenue, order count,
and average order size across four quarters of a year for five separate products.
You want to arrange these separate measures of sales in a way that enables easy
comparisons. A matrix is an ideal solution. Take a look:

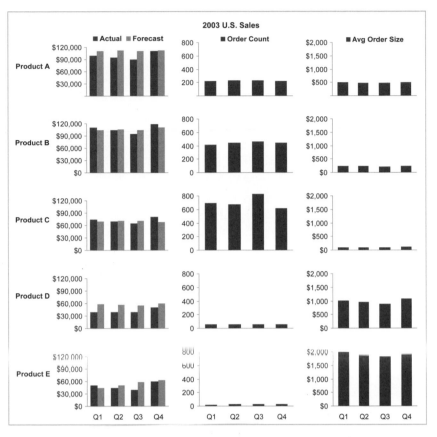

FIGURE 11.12 This matrix of graphs includes three separate but related series: forecast versus actual, order count, and average order size.

Note that a matrix of graphs that consists of multiple but related series must share the same categorical variable, which in this case is product. Otherwise, the series of graphs wouldn't be related. This type of design allows readers to examine each series independently (down the rows of a single column), examine all sales measures for a single product (across a single row), or view both at the same time looking for overall patterns. This is a powerful method for quantitative communication.

Sequence

Just like the sequencing of categorical subdivisions in a single graph, the sequencing of graphs in a series can contribute a great deal to the effectiveness of communication, especially to your readers' ability to see meaningful patterns in the information. To determine the optimal sequence of graphs, ask the two following questions:

1. Do the categorical subdivisions that vary from graph to graph (a.k.a. the *index variable*) have an intrinsic order? If so, and that order supports your message, sequence the graphs in that order. If not, proceed to the next question.
2. What quantitative value expressed in the graphs is most fundamental to the message?

The index variable (i.e., the one that varies from graph to graph) often has a built-in order that is meaningful. This order can help us determine how to sequence the graphs. For instance, if the graphs each represent data for a different year, then you will usually want to sequence the graphs chronologically. When a meaningful order is built into the index variable and there is no overriding reason to sequence the graphs in a different order, you should always sequence them accordingly. For instance, if your readers are accustomed to seeing sales regions or products in a particular order, a change in that order will seem significant, prompting them to search for its meaning. When there is no intrinsic order built into the index variable, or if it is important to your message to sequence by order of rank based on the associated quantitative values, it is time to move on to the second question above: "What quantitative value expressed in the graphs is most fundamental to the message?"

Imagine that you have a series of graphs that displays sales data, and the index variable is sales region, like the following:

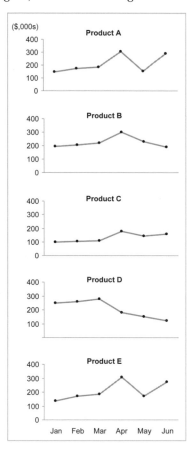

FIGURE 11.13 This is a series of related graphs that has not yet been sequenced to display a ranking relationship.

If your message focuses on the relative sales performance of the various regions, then you would want to sequence the graphs based on some measure of sales performance. If the graphs included multiple quantitative measures, such as bookings and billings, you would want to base the sequence on the one that is most significant to your message. But even if they include a single measure, as in the example above, each graph displays multiple values within that set (e.g., one for each month), so which one of those values do you base the sequence on? Take a moment to examine the previous example. Each graph displays six quantitative values: one for each month from January through June. Sequencing the graphs based on the value of a given month wouldn't give you the best result unless your objective was to rank product performance primarily for that month. How could you sequence the products based on their overall performance for the full six months?

. . . .

The answer lies in using an aggregate measure. In this case, the best choices would involve either the use of the sum or the average (mean or median) of the values. Either will give you the same result. Here's the same series once again, this time ranked in order of overall sales performance:

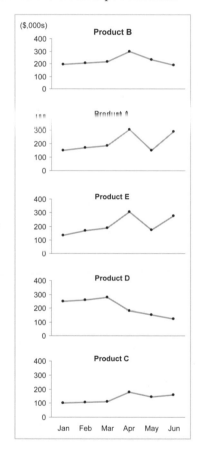

FIGURE 11.14 This is a series of related graphs that has been sequenced to display a ranking relationship.

Sums and averages aren't always the appropriate values to base your sequence on. The choice of values depends on the relationship you're displaying, the values you're using to display that relationship, and the essential point you're trying to make. If you were displaying correlations using a matrix of scatter plots, it might be appropriate to rank the graphs based on the linear correlation coefficient. If you were displaying frequency distributions, the range or median might be the best value to use for sequencing the graphs. No matter what value you use, the objective is to sequence the graphs to reveal most clearly the patterns that are central to your message.

Rules and Grid Lines

When related graphs are arranged in a series, whether horizontally, vertically, or as a matrix, rules or grid lines can be used to delineate the graphs but are rarely needed. They are only useful when one of the following circumstances exists:

- White space alone is not sufficient to delineate the graphs because of space constraints.
- The graphs are arranged in a matrix and white space alone cannot be used to direct your readers to scan the graphs in a particular sequence, either across the rows or down the columns.

The following examples illustrate how rules can be used to direct readers to scan primarily in a vertical or horizontal direction:

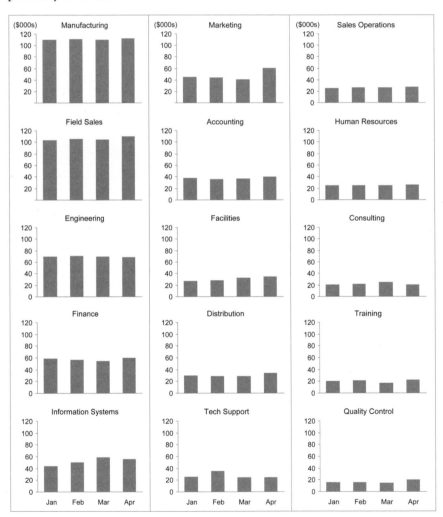

FIGURE 11.15 This matrix of graphs uses rules to direct readers to scan the graphs *vertically*.

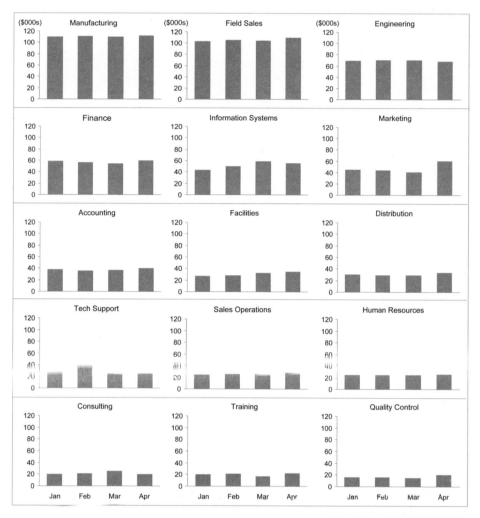

FIGURE 11.16 This matrix of graphs uses rules to direct readers to scan the graphs *horizontally*.

The vertical rules in the left-hand matrix encourage vertical scanning, and the horizontal rules in the above matrix encourage horizontal scanning. Keep in mind that rules and grid lines should always be subtly rendered, never prominently in the form of thick, dark, or bright lines.

Summary at a Glance

Combining Multiple Units of Measure

When you wish to combine two units of quantitative measure into a single graph, you may do so by placing one scale on one side of a quantitative axis and the other on the other side of that same axis.

Combining Multiple Graphs in a Series

When you need to add one more variable (i.e., another set of categorical sub-divisions) to a graph, but you've already used all the practical means to visually encode values in it, you can do so by constructing a series of related graphs, in which each graph in the series displays a different instance of the added variable.

Topic	*Practices*
Consistency	• All graphs in a related series should be consistently designed with only one exception: text used for labels, titles, or legends does not need to appear redundantly in each graph.

Arrangement	• If you want to make it easiest to compare the categorical subdivisions that vary from graph to graph:
	  • Arrange them horizontally when the quantitative scale appears on the vertical axis.
	  • Arrange them vertically when the categorical scale appears on the horizontal axis.
	• If you want to make it easiest to compare the quantitative values from graph to graph:
	  • Arrange them vertically when the quantitative scale appears on the vertical axis.
	  • Arrange them horizontally when the quantitative scale appears on the horizontal axis.
	• If there are too many graphs in the series to arrange them all vertically or horizontally on a single page or screen, arrange them in a matrix.
	• When the graphs include two quantitative scales, you may arrange the graphs horizontally, vertically, or both in the form of a matrix.
	• To display multiple measures that can't be combined in a single series of graphs, arrange them in a matrix with each series in a separate column or row.

Sequence	• If the index variable has an intrinsic order, you should sequence the graphs in this order unless you wish to display a ranking relationship. • Otherwise, rank the graphs in order based on a quantitative measure associated with the index variable.

Rules and grid lines	• Only use rules or grid lines between graphs in a series when either of these two conditions exists:
	  • The graphs must be positioned so closely together that white space alone cannot adequately delineate them.
	  • The graphs are arranged in a matrix and are positioned so closely together that white space alone cannot adequately direct your readers to scan either across or down in the manner you intend.

PRACTICE IN GRAPH DESIGN

You've come far in your exploration of graph design. It's now time for some practice to pull together and reinforce all that you've learned. Expert graph design requires that you adapt and apply what you've learned to a variety of real-world communication problems. Working through a few scenarios with a clear focus on the principles of effective graph design will strengthen your expertise and your confidence as well.

Exercise #1

I found a graph like the one pictured below in the user documentation of a popular software product that specializes in data analysis and reporting. It displays quarterly revenue and the number of guests at a particular hotel. The graph appeared as an example of the proper use of scatter plots. Because the graph isn't really designed effectively to display a correlation, and there is little value in correlating the number of guests and revenue for a hotel anyway, let's assume that its purpose is simply to display the hotel's performance in terms of revenue and number of guests by quarter for an entire year. I made no attempt while recreating this graph to make it any worse than it already was. Spend some time examining it, and then follow the instructions below.

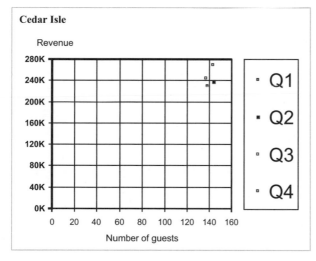

List each of the problems that you detect in the design of this graph:

Now, suggest a solution to each of these problems:

Exercise #2

This second example was derived from the user documentation of the same product that gave us the graph for Exercise #1. (I've found that I can't invent examples that are as poorly designed as those that I find quite readily in actual use.) As you can see, the intention here is to display sales revenue in the state of Kansas associated with twelve products across the four quarters of a single year. Once again, look at the graph closely and follow the instructions below.

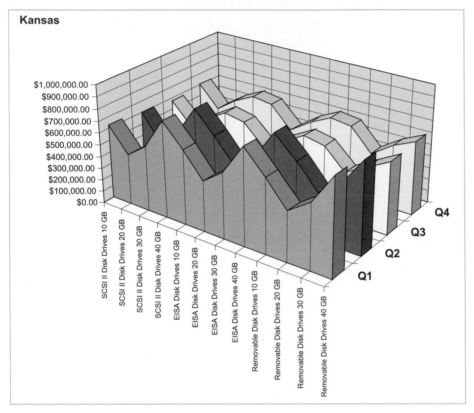

List each of the problems that you detect in the design of this graph:

Now, suggest a solution to each of these problems:

Exercise #3

The primary intention of this third example graph is to display the average selling price of gizmos as it changes monthly through the course of an entire year; the secondary intention is to relate the average selling price to the range of prices during those same months. Given these objectives, examine the graph and respond to the instructions below.

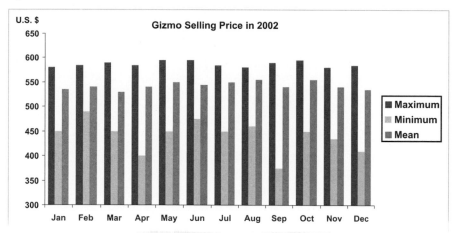

List each of the problems that you detect in the design of this graph:

Now, suggest a solution to each of these problems:

Exercise #4

It's now time to redirect your attention closer to home. Select three graphs that are used at your place of business. Make sure that at least one of them is a graph that you created. For each of the graphs, respond to the following instructions:

Graph #1

List each of the problems that you detect in the design of this graph:

Now, suggest a solution to each of these problems:

Graph #2

List each of the problems that you detect in the design of this graph:

Now, suggest a solution to each of these problems:

Graph #3

List each of the problems that you detect in the design of this graph:

Now, suggest a solution to each of these problems:

The remaining exercises ask you design graphs from scratch to achieve a specific set of communication objectives. You may construct the graphs using any relevant software that is available to you, such as spreadsheet software.

Exercise #5

Imagine that you work in the Financial Planning and Analysis department. It's your job to report how actual quarterly expenses during the previous year compared to the budget for each of the five departments that report to the Vice President (VP) of Operations. The VP is only interested in the degree to which actual expenses deviated from budget, not the actual dollar amount. Here are the raw data:

Department	Q1 Budget	Q1 Actual	Q2 Budget	Q2 Actual	Q3 Budget	Q3 Actual	Q4 Budget	Q4 Actual
Distribution	390,000	375,000	395,000	382,000	400,000	390,000	410,000	408,000
Facilities	675,000	693,000	800,000	837,000	750,000	713,000	750,000	790,000
Human Resources	350,000	346,000	350,000	342,000	350,000	340,000	350,000	367,000
Information Systems	950,000	925,000	850,000	890,000	875,000	976,000	900,000	930,000

Create one or more graphs to display this information. Once you've completed your graph, take a few minutes to describe its design, including your rationale for each design feature, in the space below.

Exercise #6

You are the company's marketing analyst. You have discovered through an analysis of last year's revenues that sales have significantly benefited from the use of television ads. You need to present this benefit to the Chief Executive Officer (CEO). Here are the data:

	Jan	Feb	Mar	Apr	May	Jun	Jul	Aug	Sep	Oct	Nov	Dec
TV Spots	20	25	20	30	30	30	30	20	15	25	30	35
Revenue	50,000	55,000	52,000	57,000	58,000	59,000	57,000	50,000	45,000	55,000	63,000	78,000

Create one or more graphs to display this information. Once you've completed your graph, take a few minutes to describe its design, including your rationale for each design feature, in the space below.

Exercise #7

For this last exercise, you are a Sales Analyst. Based on analysis of sales revenues for the past four years, you've noticed that there is a clear cyclical trend. The highest revenues are always generated during the last month of each quarter and the last quarter of each year, without exception. You want to use this information to point out the need for distributing sales activity more evenly across the year so that sales operations are not hit with such spikes in activity at quarter ends, especially at year ends. The following table contains the data, in U.S. dollars, which have not been adjusted for inflation:

Year	Jan	Feb	Mar	Apr	May	Jun	Jul	Aug	Sep	Oct	Nov	Dec
1999	50,000	52,000	55,000	51,000	52,000	55,000	48,000	49,000	50,000	53,000	56,000	62,000
2000	54,000	56,000	57,000	53,000	53,000	57,000	53,000	55,000	59,000	60,000	65,000	71,000
2001	61,000	60,000	65,000	62,000	64,000	68,000	59,000	60,000	65,000	67,000	70,000	75,000
2002	60,000	61,000	68,000	63,000	63,000	67,000	60,000	61,000	67,000	66,000	68,000	74,000

Create one or more graphs to display this information. Once you've completed your graph, take a few minutes to describe its design, including your rationale for each design feature.

You can find answers to these exercises in Appendix G, *Answers to Practice in Graph Design.*

12 THE INTERPLAY OF STANDARDS AND INNOVATION

When you design tables and graphs, you face many choices. Of the available alternatives, some are bad, some are good, some are best, and others are simply a matter of preference among equally good choices. By developing and following standards for the visual display of quantitative information, you can eliminate all but the best choices once and for all. This dramatically reduces the time it takes to produce tables and graphs as well as the time required by your readers to make good use of them. Doing this sets your skills and creativity free to be applied where they are needed most.

Ask any of my friends and they will tell you, without hesitation, that I have little respect for authority. Please don't misunderstand, I am not living under an assumed name or hiding out from the law. I simply don't like being told what to do. I prefer to examine things myself, get to know them, and then make my own informed decisions. Consequently, I'm not one who advocates rules for rules' sake. Like most professionals, I find that the term "standards" often leaves a bitter taste in my mouth. I don't like anything that gets in the way of doing good work and doing it efficiently. Nevertheless, through the course of my professional life I've come to appreciate standards when they grow out of genuine need for direction, address only what must be addressed, and save time. Standards that reside in huge tomes generally remain on the shelf gathering dust. Standards that actually get used are concise, easy to learn, easy to follow, and undeniably beneficial.

Ralph Waldo Emerson, one of the great practical philosophers of all time, wrote in his essay on self-reliance: "A foolish consistency is the hobgoblin of little minds."[1] This is often misquoted by omitting the pivotal word "foolish." Not all consistent practices are foolish. There are wise consistencies as well. These are standards that we follow because they produce the best result. These are practices that free us from redundant choices, conserving our energy and intelligence for the most deserving tasks. A good set of standards for the design of tables and graphs provides a framework for innovation in the face of important communication challenges. Standards and innovation are partners in the pursuit of excellence.

I'm amazed at how much time we waste making the same decisions over and over again simply because we've never taken the time to determine what works best for routine tasks, thereby making the decisions once and for all. Kevin Mullet and Darrel Sano beautifully express the benefits of a good set of standards:

1. Ralph Waldo Emerson (1841) *Essays: First Series, "Self-Reliance."*

Without minimizing the value of intuition as a problem solving tool,
we propose that systematic design programs are more valuable from a
communication standpoint than are ad hoc solutions; that intention is
preferable to accident; that principled rationale provides a more com-
pelling basis for design decisions than personal creative impulse.[2]

2. Kevin Mullet and Darrel Sano (1995) *Designing Visual Interfaces.* Sun Microsystems, Inc., page 15.

You will have plenty of worthwhile opportunities for creativity as you work to understand numbers and struggle to present their message in a way that leads your readers to insight and action. "Design professionals learn quickly that constraints free the designer to focus . . . resources on those portions of the problem where innovation is most likely to lead to a successful product."[3] Your time is too valuable to waste. Excite your mind with challenging and productive activity. Don't dull it with redundant, meaningless choices.

3. Kevin Mullet and Darrel Sano (1995) *Designing Visual Interfaces.* Sun Microsystems, Inc., page 227.

Not only do wise consistencies save you time and effort, they do the same for your readers. Rather than calling readers' attention to ever-changing diversity in the design of tables and graphs, you can consistently use best practices to create a calm and familiar background that invites the information itself to stand out clearly in the foreground. I appreciate the trend-setting use of simple tables and graphs that appear throughout *USA Today*, but I'm amused that each table and graph in any given issue is designed a little differently. I suppose the designers believe that such variety is necessary to keep their readers' attention. This practice does indeed attract attention, but not to the news itself.

Think for a moment about standards that you've encountered in your own work. When have they worked? When have you ignored them and why? Here are some of the qualities that design standards for tables and graphs should exhibit:

- They should grow out of real needs and never grow beyond them.
- They should represent the most effective design practices.
- They should have effective communication as their primary objective.
- They should evolve freely as conditions change or new insights are gained.
- They should be easy to learn.
- They should save time.
- They should be applied to the software that is used to produce tables and graphs in the form of specific instructions and design templates.
- They should never inhibit creativity that actually improves communication.

I won't write your design standards for you. If I did, they would likely end up gathering dust on that shelf with all the others. The standards that actually get used will be your own, the product of your own sweat, the rewards of your own labor.

Whatever you do is worth doing well. Give a clear voice to the numbers that tell the story of your business and clear visibility to the opportunities those numbers reveal. Produce tables and graphs that communicate.

Fundamental Steps in the Design Process

1. Determine your message.
2. Select the best means to display your message.
3. Design the display to *show the data*.
   - Make the data (versus non-data) prominent and clear.
   - Remove all components that aren't necessary (both data and support components).
   - Mute the remaining support components in comparison to the data.
   - Highlight above the rest the data that are most important to your message.

When to Use Tables vs. Graphs

Use Tables When	Use Graphs When
1. The document will be used to look up individual values.	1. The message is contained in the shape of the values.
2. The document will be used to compare individual values.	2. The document will be used to reveal relationships among multiple values.
3. Precise values are required.	
4. The quantitative information to be communicated involves more than one unit of measure.	

Tables: Matching Relationship and Structural Types

Quantitative-to-Categorical Relationships

	Structural Type	
Relationship	Unidirectional	Bidirectional
Between a single set of quantitative values and a single set of categorical subdivisions	Yes	Not applicable because there is only one set of categorical subdivisions
Between a single set of quantitative values and the intersection of multiple categories	Yes. Sometimes this structure is preferable because of convention.	Yes. This structure saves space.
Between a single set of quantitative values and the intersection of hierarchical categories	Yes. This structure can clearly display the hierarchical relationship by placing the separate levels of the hierarchy side by side in adjacent columns.	Yes. However, this structure does not display the hierarchy as clearly if its separate levels are split between the columns and rows.

Quantitative-to-Quantitative Relationships

	Structural Type	
Relationship	Unidirectional	Bidirectional
Among a single set of quantitative values associated with multiple categorical items	Yes	Yes. This structure works especially well because the quantitative values are arranged close together.
Among distinct sets of quantitative values associated with the same categorical item	Yes	Yes. However, this structure tends to get messy as you add separate sets of quantitative values.

Graphs: Matching Relationship Types and Value-Encoding Methods

		Value-Encoding Object		
Relationship	*Points*	*Lines*	*Points & Lines*	*Bars*
Nominal Comparison	When there is a need to narrow the quantitative scale, and in so doing, remove zero from its base	Avoid	Avoid	Either horizontal or vertical bars
Time Series	Avoid	Categorical subdivisions on X axis, quantitative values on Y axis; emphasis on overall pattern	Categorical subdivisions on X axis, quantitative values on Y axis; mutual emphasis on overall pattern and individual values	Categorical subdivisions on X axis, quantitative values on Y axis; emphasis on individual values
Ranking	When there is a need to narrow the quantitative scale, and in so doing, remove zero from its base	Avoid	Avoid	Horizontal bars are preferable, with values sorted in descending order.
Part-to-Whole	Avoid	Avoid	Avoid	Either horizontal or vertical bars
Deviation	Avoid	Especially useful when combined with time series	Useful when combined with time series and a mutual emphasis on overall pattern and individual values is desired	Either horizontal or vertical bars, except when combined with time series, which requires vertical bars
Distribution				
Single	Avoid	Known as a *frequency polygon*; emphasis on overall pattern	Avoid	Known as a *histogram*; emphasis primarily on individual values
Multiple	Use to mark the median in a box plot	Avoid	Avoid	Use in the form of range bars in box plots
Correlation	Known as a *scatter plot*	Avoid	In this case the line is a trend line, not a line that connects the points.	Either horizontal or vertical bars; can be structured either as a *correlation bar graph* or a *paired bar graph*

Appendix B   RECOMMENDED READING

EDWARD R. TUFTE is one of the world's leading theorists regarding the visual presentation of information.

The Visual Display of Quantitative Information (1983). Cheshire CT: Graphics Press.

Envisioning Information (1990). Cheshire CT: Graphics Press.

Visual Explanations (1997). Cheshire CT: Graphics Press.

WILLIAM S. CLEVELAND has written the best guidelines for sophisticated graphing techniques, especially for those used in scientific research.

The Elements of Graphing Data (1994). Summit NJ: Hobart Press.

Visualizing Data (1993). Summit NJ: Hobart Press.

NAOMI B. ROBBINS has written a book that makes many of the principles taught by Cleveland accessible to non-statisticians.

Creating More Effective Graphs (2005). New York: John Wiley & Sons, Inc.

ROBERT L. HARRIS has written a wonderful and comprehensive encyclopedia of charts in their many forms. This is a great reference to keep handy

Information Graphics (1999). New York: Oxford University Press

COLIN WARE has written the most comprehensive investigation I've found into visual perception and its application to the presentation of information. Be prepared that its scientific explanations get quite technical.

Information Visualization: Perception for Design, Second Edition (2004). San Francisco: Morgan Kaufmann Publishers.

Here are a few more related books that I keep on my shelf:

JONATHAN G. KOOMEY (2001) *Turning Numbers into Knowledge: Mastering the Art of Problem Solving.* Oakland: Analytics Press. This book provides thoughtful instruction in the use and presentation of quantitative information.

KAREN A. SCHRIVER (1997) *Dynamics in Document Design.* New York: John Wiley & Sons, Inc. This book provides thorough guidelines for general document design.

ROBERT E. HORN (1998) *Visual Language.* Bainbridge Island WA: MacroVU Press. This book provides innovative instruction in all types of visual information presentation and asserts that an entirely new visual language has emerged in recent years, which the book explores in engaging detail.

DERRICK NIEDERMAN and **DAVID BOYUM** (2003) *What the Numbers Say: A Field Guide to Mastering Our Numeric World.* New York: Broadway Books. This book provides a thoroughly readable and entertaining introduction to quantitative information as it is used in the world today.

Appendix C ADJUSTING FOR INFLATION

When the value of money decreases over time, we refer to this as *inflation*. Because this book's readers primarily reside in the United States of America, I am using U.S. dollars as the form of currency in this discussion of inflation. In order to accurately and clearly compare dollars in the past to dollars today, you must express them both using an equal measure of value. To do so you must:

- Convert today's dollars into the number equal in value to dollars in the past year that is relevant;
- Convert dollars from the past year into the number equal in value to dollars today; or
- Convert both today's dollars and the dollars from past years into the same measure of value as of some other point in time (e.g., dollars in the year 2000).

If you're not in the habit of doing this, don't get nervous. It takes a little extra work but it's not difficult. The process requires the use of an *inflation index*. Several such indexes are available, but the two that are most commonly used in the United States are the *Consumer Price Index (CPI)*, published by the *Bureau of Labor Statistics (BLS)*, which is part of the *U.S. Department of Labor*, and the *Gross Domestic Product (GDP)* deflator, published by the *Bureau of Economic Analysis (BEA)*, which is part of the *U.S. Department of Commerce*. Let's focus on the CPI as an example. The CPI represents the value of the dollar in terms of its buying power relative to goods that are typically purchased by consumers (food, utilities, etc.). CPI values are researched and computed for a variety of representative people, places, and categories of consumer goods. Let's look at a version of the CPI that represents an average of *all classes of people* across *all U.S. cities* purchasing *all types of goods* for the years 1990 through 2002:

Year	Jan	Feb	Mar	Apr	May	Jun	Jul	Aug	Sep	Oct	Nov	Dec	Annual
1990	127.4	128.0	128.7	128.9	129.2	129.9	130.4	131.6	132.7	133.5	133.8	133.8	130.7
1991	134.6	134.8	135.0	135.2	135.6	136.0	136.2	136.6	137.2	137.4	137.8	137.9	136.2
1992	138.1	138.6	139.3	139.5	139.7	140.2	140.5	140.9	141.3	141.8	142.0	141.9	140.3
1993	142.6	143.1	143.6	144.0	144.2	144.4	144.4	144.8	145.1	145.7	145.8	145.8	144.5
1994	146.2	146.7	147.2	147.4	147.5	148.0	148.4	149.0	149.4	149.5	149.7	149.7	148.2
1995	150.3	150.9	151.4	151.9	152.2	152.5	152.5	152.9	153.2	153.7	153.6	153.5	152.4
1996	154.4	154.9	155.7	156.3	156.6	156.7	157.0	157.3	157.8	158.3	158.6	158.6	156.9
1997	159.1	159.6	160.0	160.2	160.1	160.3	160.5	160.8	161.2	161.6	161.5	161.3	160.5
1998	161.6	161.9	162.2	162.5	162.8	163.0	163.2	163.4	163.6	164.0	164.0	163.9	163.0
1999	164.3	164.5	165.0	166.2	166.2	166.2	166.7	167.1	167.9	168.2	168.3	168.3	166.6
2000	168.8	169.8	171.2	171.3	171.5	172.4	172.8	172.8	173.7	174.0	174.1	174.0	172.2
2001	175.1	175.8	176.2	176.9	177.7	178.0	177.5	177.5	178.3	177.7	177.4	176.7	177.1
2002	177.1	177.8	178.8	179.8	179.8	179.9	180.1	180.7	181.0	181.3	181.3	180.9	179.9

If you need an index that focuses more directly on the value of money relative to the purchase of a specific type of item (e.g., food), by a particular class of person (e.g., clerical workers), or in a particular area of the country (e.g., the San Francisco area), it is likely that the specialized values that you need are available. You can simply go to the *Bureau of Labor Statistics (BLS)* website and select what you need from the broad range of available data. It is very easy to transfer the data from the website directly to software on your own computer, such as *Microsoft Excel* spreadsheet software. In fact, I was able to get the information for this sample table simply by electronically copying it from the website and pasting it into a spreadsheet.

Once you have an index in a form that is readily available for your use, such as a spreadsheet, here's how you actually use it. The current version of the CPI uses the value of dollars from 1982 to 1984 as its baseline. So each value in the index represents the value of dollars at that time compared to their value from 1982 to 1984. For instance, according to the above table, in January of 1990, the value of the dollar was 127.4% of its value in 1982 to 1984, and for the year 1990 as a whole, it was 130.7% of its value in 1982 to 1984. Typically, if you were comparing money across a range of time, you would express everything according to the value of money at some point along that range, and most often in terms of its value at the point in the range that represents the value of money at the time you were producing your report. If you were producing your report in 2002, including values ranging from 1998 to 2002, you would likely want to convert all the values to their 2002 equivalent. Here's how you would convert a year 1998 value of $100,000 into its year 2002 equivalent, assuming that you are only dealing with one value per year (e.g., as opposed to monthly or quarterly values).

1. Find the index value for the year 2002, which is 179.9.
2. Find the index value for the year 1998, which is 163.0.
3. Divide the index value for 2002 by the index value for 1998, which results in 1.103681.
4. Multiply the dollar value for 1998, which is $100,000, by the results of step 3, which is 1.103681, which results in $110,368.10, which you can round to the nearest whole dollar, reaching the final result of $110,368.

Because the year 2002 dollars are already expressed as 2002 dollars, you don't have to convert them. If you're using spreadsheet software, setting up the formulas to convert money using an inflation index like the CPI is quite easy to do.

Whether you decide to express money across time using an inflation index to convert it to a common base or to use the actual values without adjusting them for inflation, you should always clearly indicate on your report what you've done. Don't leave your readers guessing. As a communicator of important information, labeling the way that you have expressed the value of money is a practice that you should get into the habit of following. If you haven't adjusted for inflation, you can simply state somewhere on the report that you are using *Current U.S. Dollars*. If you have adjusted for inflation, use a statement like *Adjusted to a base of year 2002 U.S. dollars* or *Adjusted according to the CPI using a baseline of year 2002*.

For additional information on this topic, along with comprehensive instruction in the use of quantitative information, I recommend that you get a copy of Jonathan Koomey's excellent book *Turning Numbers into Knowledge*, published by Analytics Press.

Appendix D   CONSTRUCTING CORRELATION BAR AND PAIRED BAR GRAPHS WITH MICROSOFT EXCEL

Constructing Paired Bar Graphs

1. Sort the data sets by the size of one of the variables.

2. Create a column graph. First, define a data series for the salaries; then, define a data series for height.

3. Only the salary series will be visible for now, with its scale on the Y axis on the left. Select the salary series by clicking one of its bars; then, right click and select *Format data series*. Select the *Axis* tab and change the *Plot series on* selection to *Secondary axis*. This will place the salary scale on a new Y axis on the right, revealing the height scale on the Y axis on the left. While still in the formatting window, select the *Options* tab and set the *Gap width* to around 150 to increase the gap between the bars, thereby narrowing the bars. Next select the height data series by clicking one of its bars, and right click in order to select *Format data series*. Select the *Options* tab and decrease the *Gap width* to around 30 to reduce the gap between bars, thereby widening the bars.

4. Select each data series individually, then right click and select *Add trend line*.

Constructing Correlation Bar Graphs

1. Sort the data sets by the size of one of the variables.

2. Create a bar graph for the unsorted data set. Remove the category axis. Create a title for the data set.

3. Copy this graph and place the copy to the immediate left of the first version.

4. In the graph on the left, change the source data to the sorted data set. Select the Y axis, then right click with the mouse to get the list of possible actions and select *Format Axis*. Select the *Scale* tab and check the *Values in reverse order* box.

Appendix E ANSWERS TO PRACTICE IN SELECTING TABLES AND GRAPHS

Scenario #1

1. **Table or graph?** Because the CFO needs a simple reference that contains budgeted and actual headcount and expense figures for the current quarter so that she can quickly locate the numbers for each of her meetings, the report will be used primarily to look up data. Therefore, a table will work most effectively.

2. **If a table, which kind?** This table should take the form of a list, most likely with a row for each department and separate columns for the four sets of quantitative values.

3. **If a graph, what kind of relationship will it display?** Not applicable.

4. **If a graph, which graphical object or objects will you use to encode the quantitative values?** Not applicable.

5. **Is there anything else you would be sure to include in this presentation?** ⟨illegible⟩

Scenario #2

1. **Table or graph?** Because you need to show a pattern rather than individual values, a graph will work best.

2. **If a table, which kind?** Not applicable.

3. **If a graph, what kind of relationship will it display?** This graph needs to display two relationships of equal significance to your message: 1) a part-to-whole relationship, comparing the relative revenue contributions to total revenue of each of the product lines, and 2) a time-series relationship, showing how the relative contributions of the individual product lines have changed during the past five years.

4. **If a graph, which graphical object or objects will you use to encode the quantitative values?** Although part-to-whole relationships are usually encoded as bars, we need to show five sets of values, one for each product line, so the trend of revenue contribution across time would be difficult to see using five sets of five bars, with a distinct bar for each product line and a set of bars for each year. A distinct line on the graph to represent each of the product lines, each line a different color, will display the trend and relative values of the product lines more clearly.

5. **Is there anything else you would be sure to include in this presentation?** To clearly emphasize the decline over time of the product lines that were on top five years ago, it would help to sequence the five product lines in order of revenue contribution five years ago. Also, since you want to emphasize the decline of the programming and utilities product lines, you might want to make their lines stand out somewhat from the others by increasing their thickness.

Scenario #3

1. **Table or graph?** If your director was comfortable interpreting measures of average like means and medians and measures of range like standard deviations, you could provide him with a very simple table containing the median, mean, and standard deviation for each of the two classes. Because he's not comfortable with standard deviations, you should use an approach that displays the information graphically.

2. **If a table, which kind?** Not applicable.

3. **If a graph, what kind of relationship will it display?** The median and the distribution of the ratings across the five values (1 through 5) can both be displayed by means of a frequency distribution.

4. **If a graph, which graphical object or objects will you use to encode the quantitative values?** Because you want to highlight the median value and the range of values, a histogram using vertical bars will display both quite effectively. The histogram would consist of five bars, one bar for each rating value of 1 through 5.

5. **Is there anything else you would be sure to include in this presentation?** Because you want to compare the ratings of two different courses, you could switch to a frequency polygon, using a separate line to represent the frequency distribution across the ratings for each course, but in doing so you would lose some of the visual emphasis on the median value of each. A better solution involves two histograms: one for each course, positioned one above the other so that it is easy to compare their medians and ranges of distribution. If you want to include the arithmetic means as well, you could mark them on the X axes of the graphs or draw a vertical line in each graph where the mean value falls along the X axis.

Scenario #4

1. **Table or graph?** Even though you will be communicating a ranking of the four customer service centers to point out the ones that need the most improvement, your message consists of only four values. Given such a small data set, this message can be communicated more efficiently using a table than a graph, without losing any meaning.

2. **If a table, which kind?** This is a simple list, with one value for each customer service center.

3 **If a graph, what kind of relationship will it display?** Not applicable.

4. **If a graph, which graphical object or objects will you use to encode the quantitative values?** Not applicable.

5. **Is there anything else you would be sure to include in this presentation?** The message will be clearest if you sort the customer service centers in the table in order of their average rating.

Scenario #5

1. **Table or graph?** You want to show a pattern, so a graph will definitely work the best.
2. **If a table, which kind?** Not applicable.
3   **If a graph, what kind of relationship will it display?** It will display a correlation between the number of workers and their productivity.
4. **If a graph, which graphical object or objects will you use to encode the quantitative values?** If the Operations Manager is familiar with scatter plots, you could use this form of display, with points for each monthly measure of headcount and productivity, and a trend line to highlight the strong negative correlation. If he is not, you could use a combination of bars and trend lines in the form of either a correlation bar graph or a paired bar graph. Even if he understands scatter plots, a scatter plot will not highlight the fact that there was a particular point in time after his arrival when the negative correlation began to appear. A correlation bar graph, the months labeled along the X axis, headcount on the left Y axis, and productivity on the right Y axis, will display this with glaring clarity.
5. **Is there anything else you would be sure to include in this presentation?** Perhaps an arrow pointing to the month when the hiring of additional workers began, distinctly labeled, would add further clarity.

Scenario #6

1. **Table or graph?** A graph will highlight the dramatic nature of this information more effectively than a table would. The quick fall-off of the revenue contribution of your company's orders from 87% for the top 10% of the orders to 1% for the bottom 50% lends itself to a visual display.
2. **If a table, which kind?** Not applicable.
3. **If a graph, what kind of relationship will it display?** This scenario is a little tricky. Essentially, the information involves a part-to-whole relationship: the revenue contribution of each 10% group of orders as a percentage of total revenue. Even though our separation of the orders into 10 groups involved a ranking of them based on their revenue amount, the point of our message is the contribution of our biggest orders to overall revenue, compared to the contributions of the other groups of orders of decreasing size.
4. **If a graph, which graphical object or objects will you use to encode the quantitative values?** A part-to-whole relationship is most effectively displayed using bars. In this case each bar represents the percentage revenue contribution of each 10% group of orders.
5. **Is there anything else you would be sure to include in this presentation?** Very clear labeling along both axes is necessary to avoid confusion between the 10% groups of orders vs. the percentage contribution of each group to overall revenue. Also, a title that clearly and succinctly states the essential message, such as "The Top 10% of Our Orders Account for 87% of Our Revenue" would catch the reader's attention, allowing the graph to then dramatically reinforce the message.

Appendix F   ANSWERS TO PRACTICE IN TABLE DESIGN

Exercise #1

Here's the original table:

Quarter-to-Date Sales Rep Performance Summary
Quarter 2, 2003 as of March 15, 2003

Sales Rep	Quota	Variance to Quota	% of Quota	Forecast	Actual Bookings
Albright, Gary	200,000	-16,062	92	205,000	183,938
Brown, Sheryll	150,000	84,983	157	260,000	234,983
Cartwright, Bonnie	100,000	-56,125	44	50,000	43,875
Caruthers, Michael	300,000	-25,125	92	324,000	274,875
Garibaldi, John	250,000	143,774	158	410,000	393,774
Girard, Jean	75,000	-48,117	36	50,000	26,883
Jone, Suzanne	140,000	-5,204	96	149,000	134,796
Larson, Terri	350,000	238,388	168	600,000	588,388
LeShan, George	200,000	-75,126	62	132,000	124,874
Levensen, Bernard	175,000	-9,267	95	193,000	165,733
Mulligan, Robert	225,000	34,383	115	275,000	259,383
Tetracelli, Sheila	50,000	-1,263	97	50,000	48,737
Woytisek, Gillian	190,000	-3,648	98	210,000	186,352

List each of the problems that you detect in the design of this table:

1. The grid that delineates the columns and rows is far too heavy and isn't necessary at all.

2. The combined use of boldfacing and the gray fill color to delineate the headers is excessive, grabbing too much attention relative to the values.

3. The center alignment of the headers fails to preview the alignment of their values.

4. The values in the *% of Quota* column aren't clear without percentage signs.

5. Nothing in the columns that contain dollar values indicates the unit of measure.

6. Overall performance for the entire group cannot be determined without summing the values in the columns.

7. Nothing has been done to make it easy to compare the performance of the individual sales representatives.

8. The most important and useful set of quantitative values is bookings, yet they are in the last column, furthest from the names of the sales representatives, causing a great deal of eye movement to pair them.

9. The *Variance to Quota* and *% of Quota* columns appear before the *Bookings* column, which is used to calculate them.

Now, suggest a solution to each of these problems:

1. Eliminate the grid entirely, using only white space to delineate the columns and the rows.

2. Set the headers apart from the values using a rule line and nothing more. The headers could be boldfaced to set them apart further, but this really isn't necessary and would result in making them more prominent than the values they describe.

3. Align the headers consistently with the values they describe.

4. Format the values in the percentage column with percentage signs.

5. Format the summary values in the columns containing dollars to clarify the unit of measure, assuming that the context does not require further clarification to indicate U.S. dollars, Canadian dollars, etc.

6. Add column summaries to provide measures of overall performance.

7. To enhance comparison of sales representative performance, sort the rows by bookings and add a column that expresses the bookings of each sales representative as a percentage of total bookings.

8. Place the column of bookings values immediately to the right of the sales representatives' names.

9. Place the calculated columns containing the bookings-to-quota variance and bookings percentage-of-quota values to the right of both the bookings and the quota columns.

Here's an example of what the table might look like when designed effectively.

Quarter-to-Date Sales Rep Performance Summary
Quarter 2, 2003 as of March 15, 2003

Sales Rep	Actual Bookings	% of Total Bookings	Forecasted Bookings	Quota	Bookings to Quota Variance	Bookings % of Quota
Larson, Terri	588,388	22.1%	600,000	350,000	238,388	168%
Garibaldi, John	393,774	14.8%	410,000	250,000	143,774	158%
Caruthers, Michael	274,875	10.3%	324,000	300,000	-25,125	92%
Mulligan, Robert	259,383	9.7%	275,000	225,000	34,383	115%
Brown, Sheryll	234,983	8.8%	260,000	150,000	84,983	157%
Woytisek, Gillian	186,352	7.0%	210,000	190,000	-3,648	98%
Albright, Gary	183,938	6.9%	205,000	200,000	-16,062	92%
Levensen, Bernard	165,733	6.2%	193,000	175,000	-9,267	95%
Jone, Suzanne	134,796	5.1%	149,000	140,000	-5,204	96%
LeShan, George	124,874	4.7%	132,000	200,000	-75,126	62%
Tetracelli, Sheila	48,737	1.8%	50,000	50,000	-1,263	97%
Cartwright, Bonnie	43,875	1.6%	50,000	100,000	-56,125	44%
Girard, Jean	26,883	1.0%	50,000	75,000	-48,117	36%
Total	$2,666,591	100.0%	$2,908,000	$2,405,000	$261,591	111%

Exercise #2

Here's the original table:

Mortgage Loan Rates
Effective September 1, 2003

Loan Type	Term	Points	Lender	Rate
Adjustable	15	0	ABC Mortgage	6.0%
Adjustable	15	0	BCD Mortgage	6.0%
Adjustable	15	0	CDE Mortgage	6.0%
Fixed	15	0	ABC Mortgage	6.25%
Fixed	15	0	BCD Mortgage	6.75%
Fixed	15	0	CDE Mortgage	7.0%
Adjustable	30	.5	ABC Mortgage	6.125%
Adjustable	30	.5	BCD Mortgage	6.25%
Adjustable	30	.5	CDE Mortgage	6.5%
Fixed	30	.5	ABC Mortgage	6.5%
Fixed	30	.5	BCD Mortgage	7.0%
Fixed	30	.5	CDE Mortgage	7.25%
Adjustable	15	1	ABC Mortgage	5.675%
Adjustable	15	1	BCD Mortgage	5.675%
Adjustable	15	1	CDE Mortgage	5.75%
Fixed	30	1	ABC Mortgage	6.5%
Fixed	30	1	BCD Mortgage	6.5%
Fixed	30	1	CDE Mortgage	7.0%
Adjustable	15	1	ABC Mortgage	5.675%
Adjustable	15	1	BCD Mortgage	5.675%

List each of the problems that you detect in the design of this table:

1. The information is not grouped and sorted properly for looking up all the rates for a single lender.
2. The *Points* and the *Lender* columns are not adequately delineated. There is not enough white space between to view them separately without effort.
3. The fluctuating precision of the values in the *Points* and *Rate* columns makes them difficult to read when scanning vertically.
4. The row headers (i.e., values in the *Loan Type*, *Term*, and *Points* columns) are unnecessarily repeated on each row, making it difficult to see when one group ends and the next begins.

Now, suggest a solution to each of these problems:

1. Place the lender names in the first column, followed by the other categorical values, with the rates in the right column.
2. Sometimes when a column with right-aligned values is positioned immediately to the left of a column with left-aligned values, there isn't enough white space between them to separate them adequately. This can be corrected with most software programs by placing a blank column between them in order to insert more white space.
3. Select the appropriate numeric precision; then, format all values to display this level of precision.
4. When rows are grouped by categorical values that appear to the left of the quantitative values, you only need to display those values on the first row of the group and on the first row of each new page that contains the group.

Here's an example of what the table might look like when designed effectively:

Mortgage Loan Rates by Lender Effective September 1, 2003				
Lender	Loan Type	Term	Points	Rate
ABC Mortgage	Adjustable	15 Yr	0.0	6.000%
			0.5	5.750%
			1.0	5.675%
		30 Yr	0.0	6.125%
			0.5	5.875%
			1.0	5.500%
	Fixed	15 Yr	0.0	6.250%
			0.5	6.000%
			1.0	5.750%
		30 Yr	0.0	6.500%
			0.5	6.000%
			1.0	5.750%
BCD Mortgage	Adjustable	15 Yr	0.0	6.000%
			0.5	5.750%
			1.0	5.675%
		30 Yr	0.0	6.125%

Note that in this example, the lenders' names have been highlighted using bold-facing. This isn't necessary, but because the purpose of the table is to look up specific lenders, it is useful to make the names stand out visually from the other data.

Exercise #3

Here's the original table:

2003 Marketing Department Expenses			
Quarter	Transaction Date	Expense Type	Expense
	9/28/2003	Software	3837.05
	9/28/2003	Computer Hardware	10873.34
	9/29/2003	Travel	2939.95
	9/30/2003	Supplies	27.53
Qtr 4	10/1/2003	Supplies	17.37
	10/1/2003	Postage	23.83
	10/3/2003	Computer Hardware	3948.85
	10/3/2003	Software	555.98
	10/3/2003	Furniture	739.37
	10/3/2003	Travel	28.83
	10/4/2003	Entertainment	173.91
	10/15/2003	Travel	33.57
	10/16/2003	Membership Fees	395.93
	10/16/2003	Conference Registration	2195.00

List each of the problems that you detect in the design of this table:

1. The information is not conveniently arranged for the examination of expenses by expense type per quarter. The current arrangement would require a great deal of searching and summing to generate the needed information.

2. The table contains values that aren't necessary and that consequently just get in the way. Because the marketing manager doesn't need date information below the quarter level, the transaction dates are not only unnecessary, they force the information down to a level of detail that increases its volume considerably.

3. Highlighting the quarters, expense types, and expense amounts through the use of boldfacing is unnecessary. Everything but the transaction dates has been highlighted to make the dates appear less important, but the dates are totally unnecessary and therefore shouldn't be included at all.

4. If the dates were needed, their alignment to the right and their fluctuating number of digits would render them inefficient to read.

5. Summary values at the required level of quarter and expense type are missing.

6. The expense amounts are missing the comma used to group every three whole number digits, making them harder than necessary to read.

7. Tying expense types and expenses closely together, if it were useful, could have been achieved through a better means of visual grouping. Even if a distinct font were appropriate, the font that was used in this table is not very legible.

8. The heavy vertical rules lead the eyes to scan downward through the columns, yet horizontal scanning between the expense types and expenses is the primary way that this table should be read.

9. Failing to repeat the quarter row header at the top of each new page could result in pages that have no row header at all, therefore forcing readers to flip backward to previous pages to determine the quarter.

Now, suggest a solution to each of these problems:

1. Arrange the expense types in a column along the left with individual quarters across the top; this arrangement nicely supports the need to examine expenses by type in total and by quarter.

2. Eliminate the transaction dates entirely, displaying expenses only at the level of expense type and quarter.

3. By eliminating the transaction dates you eliminate the need to make the other values stand out in relation to them. It would be useful, though, to highlight total expenses by expense type as distinct and somewhat more important than the quarterly expenses.

4. Date alignment and formatting are no longer a problem because the dates have been eliminated.

5. When you eliminate the transaction dates, all quantitative values will automatically be summarized at the appropriate level of expense type by quarter. Total expenses by expense type can be added as a separate column.

6. Expense value formats can be corrected by adding the commas to group every three whole-number digits.

7. Eliminate the distinct font because it serves no useful purpose.

8. Eliminate the vertical rules so that horizontal scanning is not interrupted.

9. The elimination of the transaction dates eliminates the need to repeat row headers on each new page because there is now only one row per row header.

Here's an example of what the table might look like when designed effectively:

2003 Marketing Department Expenses

Expense Type	Total	Qtr 1	Qtr 2	Qtr 3	Qtr 4
Computer Hardware	12,970.40	3,883.64	8,352.83	0.00	733.93
Entertainment	4,778.52	736.94	1,873.03	185.01	1,983.54
Equipment	463.79	73.93	105.93	283.93	0.00
Furniture	2,541.11	0.00	108.83	493.83	1,938.45
Membership Fees	1,595.00	95.00	1,500.00	0.00	0.00
Postage	292.59	27.83	186.83	57.93	20.00
Supplies	444.06	117.93	75.39	74.82	175.92
Travel	12,674.09	5,938.93	3,978.39	863.02	1,893.75

Note that the *Total* column has been placed immediately to the right of the *Expense Type* column, making it very easy and efficient to associate the two sets of values. Because total expenses per expense type are somewhat more useful than any individual column of quarterly expenses, this arrangement of the columns makes the total slightly easier to read.

Exercise #4

Answers don't apply to this exercise.

Exercise #5

Here's one potential design solution to the proposed scenario:

YTD Product Performance through 5/31/03

Our top 2 products account for more than 89% of revenue and 95% of profit!

Product	Revenue			Profit		
	Dollars (000s)	% of Total	Cum % of Total	Dollars (000s)	% of Total	Cum % of Total
I	9,266	74.01%	74.01%	5,969	79.15%	79.15%
F	1,957	15.63%	89.64%	1,197	15.87%	95.02%
G	602	4.81%	94.45%	207	2.75%	97.77%
E	402	3.21%	97.66%	15	0.20%	97.97%
B	132	1.05%	98.71%	73	0.96%	98.93%
D	92	0.74%	99.45%	39	0.51%	99.44%
C	40	0.32%	99.77%	32	0.43%	99.88%
H	20	0.16%	99.93%	5	0.07%	99.94%
A	7	0.05%	99.98%	4	0.05%	99.99%
J	2	0.02%	100.00%	1	0.01%	100.00%
Total	$12,520	100.00%	100.00%	$7,541	100.00%	100.00%

Some of the highlights of this solution are:

1. The values have been sorted in descending order by revenue, thus placing the products in order of performance, beginning with the best. The values could also have been sorted by profit or by a combination of revenue and profit. Given this arrangement of values, with revenue first, then profit, it was appropriate to sort by revenue.

2. Because the only measures of performance that need to be communicated are revenue and profit, other measures that might have been included (e.g., cost or units sold) are left out.

3. The message that the *top 2* products performed well above the rest has been highlighted in multiple ways: 1) boldfacing these two rows, 2) placing a concise textual statement of this fact in red above the table, and 3) placing red borders around the two measures that most directly present the contribution of these two products to overall revenue and profit.

4. Additional measures were calculated and included to enhance the meaning and clarity of the relative revenue and profit contributions of each product, including percent of total and cumulative percent of total. These measures make the limited contribution of the worst performing products stand out even more than the dollars alone.

5. The title clearly states the period of time covered by the performance figures. Without this information, the meaning of the table would suffer even in the beginning and would quickly decrease in worth with the passage of time.

The design in this example solution is not the only design that would work effectively. Even if your design is significantly different, compare its merits to those of the example table to see how well it did in meeting the objectives specified in the scenario.

Appendix G ANSWERS TO PRACTICE IN GRAPH DESIGN

Exercise #1

Here's the original graph:

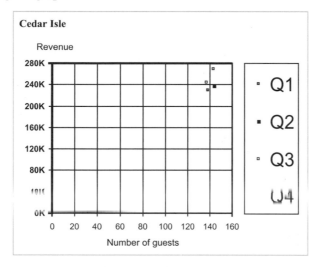

List each of the problems that you detect in the design of this graph:

1. The fundamental problem with this graph is that it is the wrong type. Scatter plots display correlations, but the message here does not involve a correlation. Rather, it is a time-series relationship that involves two sets of quantitative values expressed in two different units of measure (revenue in dollars and number of guests as a count).
2. The four data points are too small to be easily distinguishable.
3. The grid lines are not necessary, and, even if they were, they are too visually prominent.
4. The axes are too visually prominent, standing out more than the data objects.
5. There is no reason for the labels along the vertical scale to be visually prominent.
6. There is no reason for the legend containing the quarters to be visually prominent, including its enclosure in a border and much bigger font than anything else.
7. The title of the vertical axis (i.e., *Revenue*) is not aligned with the labels.
8. The title of the graph (i.e., *Cedar Isle*) uses a different font from the rest for no apparent reason.
9. There is no date or other indicator of the year from which the data were taken or any indicator of the currency.

Now, suggest a solution to each of these problems:

1. Use three line graphs, arranged vertically, with quarters along the horizontal axis: one with revenue dollars, one with the number of guests, and one with the average revenue per guest in dollars along their vertical axes.
2. If points are used to encode the data, make them larger.
3. Eliminate the grid lines.
4. Visually mute the axis lines compared to the data.
5. Make the revenue dollar text labels no more visually prominent than the other labels.
6. Eliminate the legend.
7. Align the axis titles with the labels.
8. Use the same font throughout the graph.
9. Add a date and an indicator of the currency.

Here's an example of what the solution might look like when designed effectively:

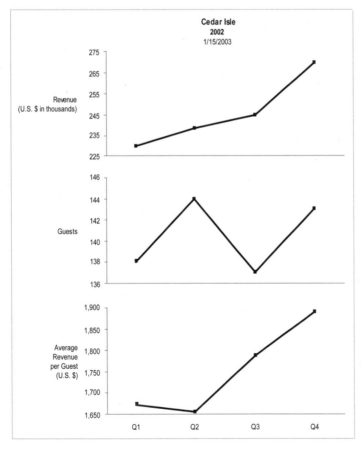

Exercise #2

Here's the original graph:

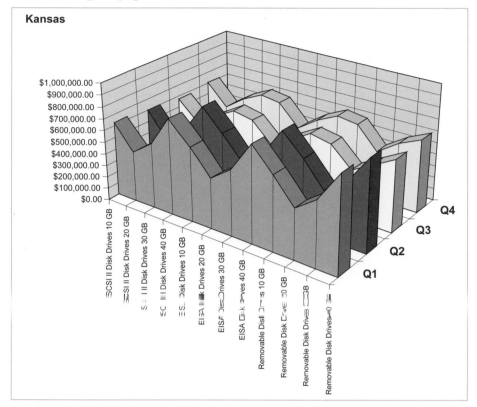

List each of the problems that you detect in the design of this graph:

1. The most significant problem with this graph is that lines have been used to connect values that are discrete, with no intimate connection. It is not meaningful to connect values along a nominal or ordinal scale, but only along an interval scale.

2. This graph is 3-D, which makes it impossible to read.

3. Encoding the values as three-dimensional objects also makes them more difficult to read.

4. The graph contains too many categorical items. It is difficult to display twelve products by four quarters in a single graph.

5. There are too many tick marks and labels along the vertical axis.

6. The numbers along the vertical axis should not include decimals. Including them makes the numbers less efficient to read and suggests a level of numeric precision that isn't available.

7. The labels for the quarters do not need to be more prominent than the other sets of labels.

8. The vertical orientation of the labels along the horizontal axis makes them difficult to read.

9. The fill pattern on the side and back walls adds no value but certainly adds distraction.

10. The grid lines add no value.

11. There is no indication of the year or the currency type.

Now, suggest a solution to each of these problems:

1. Restrict the graph to two dimensions.
2. Encode this time-series data as separate lines for each product.
3. Construct a series of graphs rather than trying to encode all 12 products in a single graph. Because the products group nicely into three sets of four products each (SCSI II, EISA, and Removable disk drives), each of these groups can be displayed in a single graph to make comparisons easier.
4. Reduce the number of tick marks and labels along the quantitative scale.
5. Remove the decimals from the numbers on the quantitative scale.
6. Make all the scale line labels equal in visual prominence.
7. Remove the products from the X axis, which eliminates the need to change their orientation.
8. Eliminate the third dimension, which removes the walls, eliminating the need to remove the distracting fill pattern.
9. Remove the grid lines.
10. Add the year to the graph title and the currency type to the Y axis title.

Here's an example of what the graph might look like when designed effectively:

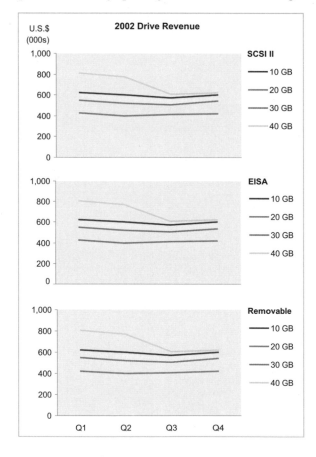

Note that the data regions have been slightly highlighted with a light gray fill in order to draw focus to them and make them distinct from the surrounding objects (e.g., the legend).

Exercise #3

Here's the original graph:

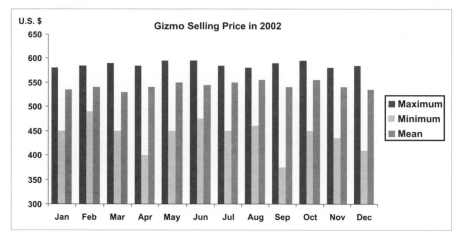

List each of the problems that you detect in the design of this graph:

1. Using separate bars to encode the monthly maximum, minimum, and mean values is not suitable for visualizing the trend of the mean through time or its relationship to the full range of monthly values.
2. The most important set of values is the means, yet the means are less visually prominent than the maximum values.
3. Sequencing the mean values after the minimum values is awkward. It is more natural to position the means between the maximum and minimum values.
4. Far too many components of this graph are visually prominent. It is hard to look at. The data do not stand out. The most visually prominent component of the entire graph is the legend because of its enclosure within borders.
5. The tick marks along the vertical axis are not necessary.

Now, suggest a solution to each of these problems:

1. The trend of the mean values as they march through time should be encoded as a line. The maximum and minimum values may also be encoded as lines or as range bars.
2. Highlight the line that encodes the mean values.
3. Using lines or range bars to encode the maximum and minimum values eliminates the problem associated with the awkward sequence of the bars.
4. Visually mute the supporting components.
5. Eliminate the tick marks along the categorical axis.

Here's an example of what the graph might look like when designed effectively:

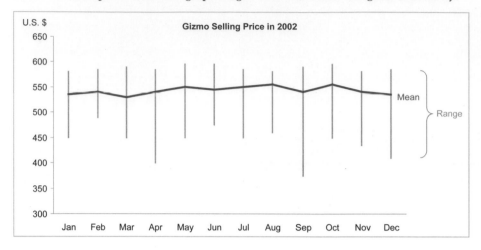

Exercise #4

Answers don't apply to this exercise.

Exercise #5

Here's one potential design solution for the proposed scenario:

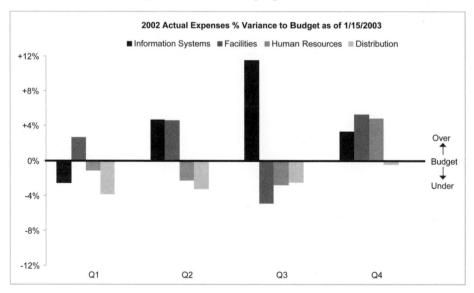

Some of the highlights of this solution are:

1. Because the purpose of this graph is to display a deviation relationship, it is ideal to convert the actual dollar expenses to a percentage difference from the budget. Setting the baseline of the quantitative scale to the budget value of 0% makes deviations easy to see as bars extending up or down from that baseline.

2. The values could have been encoded as lines rather than bars, but the bars place more emphasis on the distinct quarterly values rather than the trend over time.

The design in this example solution is not the only design that would work effectively. Even if your design is significantly different, compare its merits to those of the example solution to see how well it did in meeting the objectives specified in the scenario.

Exercise #6

Here's one potential design solution for the proposed scenario:

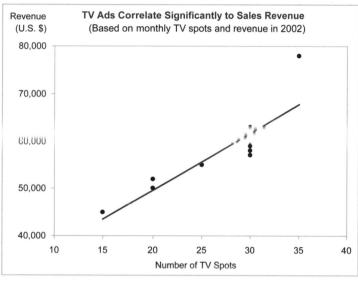

Some of the highlights of this solution are:

1. Because the purpose of this graph is to display a correlation between two variables, not to track the values across time, a scatter plot is an ideal choice.
2. The use of a trend line makes the strong positive correlation obvious.
3. The use of the contrasting red hue for the title and the trend line draws readers' eyes to the two most important pieces of information: the conclusion of the analysis and the data that support that conclusion.
4. Subtle borders around the data region form a useful enclosure to keep the residual value of $78,000 in the upper right corner from seeming extraneous.

Exercise #7

Here's one potential design solution for the proposed scenario:

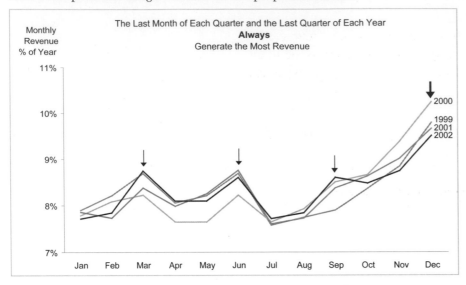

Some of the highlights of this solution are:

1. Because the message requires that the same months of each of the four years be compared to one another, the most effective layout involves assigning each of the years to the same set of twelve months rather than arranging the four years in sequence from left to right along the horizontal axis.

2. Rather than adjusting the revenue numbers for inflation, I've converted each month's revenue into a percentage of the entire year's revenue, which eliminates the effect of inflation.

3. A clear title, which points out the important message contained in the data, makes the graph very easy to interpret.

4. The arrows clearly point out the last month of each quarter and the last quarter of each year, thereby visually reinforcing the message.

5. Labeling the lines directly eliminates the need for a legend.

Index

About the Author . . .

STEPHEN FEW has over 20 years of experience as an innovator, consultant, and educator in the fields of *data warehousing* (a.k.a. *business intelligence* and *decision support*) and *information design.* Today, as the founder and principal of the consulting firm **Perceptual Edge**, Stephen focuses on the design of information for effective communication. When he isn't working, he can normally be found—in or around his home in Berkeley—lost in a good book, savoring a delightful wine, hiking in the hills, or instigating an animated discussion about the meaning of life with good friends.